Audel™

Managing Maintenance Planning and Scheduling

Michael V. Brown

WILEY

Wiley Publishing, Inc.

The author would like to acknowledge the contributions of Michael H. Bos, cofounder and former partner in New Standard Institute, in the development of the original version of this material.

Vice President and Executive Group Publisher: Richard Swadley
Vice President and Executive Publisher: Robert Ipsen
Vice President and Publisher: Joseph B. Wikert
Executive Editor: Carol A. Long
Editorial Manager: Kathryn A. Malm
Senior Production Manager: Fred Bernardi
Development Editor: Kevin Shafer
Production Editor: Vincent Kunkemueller
Text Design & Composition: TechBooks

Library of Congress Cataloging-in-Publication Data:

Brown, Michael, 1951–
 Audel managing maintenance planning and scheduling / Michael Brown.
  p. cm.
Includes index.
ISBN: 0-7645-5765-3 (paper/website)
1. Plant maintenance—Management. 2. Plant maintenance—Planning.
3. Business logistics. I. Title
 TS192.B75 2004
 658.2′02—dc22

                              2004005522

Printed in the United States of America

10 9 8 7 6 5 4 3 2 1

Contents

About the Author

Michael V. Brown is an electrical and maintenance engineer with 30 years of experience in industry, and has held positions at both the plant and corporate levels of Fortune 500 companies. As founding partner and president of New Standard Institute, he has designed and implemented maintenance management improvement programs for numerous industrial clients. New Standard Institute provides seminars, consultations, and computer-based training programs specific to maintenance-related subjects. For more than a decade, he has written articles that have been published in many U.S. and Canadian magazines, as well as on the Internet. Other books he has written and published by Wiley include *Audel Managing Maintenance Storerooms* and *Audel Managing Shutdowns, Turnarounds, and Outages.*

Introduction

I have to admit that it was not my original intention to enter the maintenance profession. I first got an associates degree and went to work in the metrology lab at Pratt and Whitney (a designer and manufacturer of turbine engines). The position was really a maintenance job—calibrating and repairing instrumentation in engine test cells. After a few years, I went back to school to pursue a bachelor's degree in electrical engineering. My studies focused on electronics and the burgeoning field of microcomputers, and I wanted to be part of it. (Intel had created the world's first microprocessor just a few years earlier.) But this was not to be. The Vietnam War was winding down when I finally got my degree. Experienced electronic engineers in the defense industry were losing their jobs by the thousands. It was hard to find a job anywhere, never mind an engineering job. I finally landed a job with Monsanto as a maintenance engineer, and I was glad to get it.

In the ensuing years, I recognized the professionalism that exists at all levels in a maintenance organization. No one was just interested in solving the day-to-day breakdowns. Everyone was conscious of the need to find long-term solutions to problems. It is always a struggle to focus on one while working on the other.

Later maintenance positions within Olin Corporation and Stauffer Chemical solidified this impression of maintenance professionalism. At Stauffer I was in a corporate position, assisting plants in improving their maintenance effort. This afforded me the opportunity I seldom had while working at the plant level: determining the long-term solutions to maintenance problems. The real answer always came down to planning and scheduling the maintenance work. Maintenance workers want to be gainfully employed. They really don't like waiting for instructions from their supervisors for every move they make, waiting for parts at the storeroom counter, waiting for another craft to complete the next phase of a job, or traveling back and forth for needed equipment and parts. They would prefer to work for an organization that has a direction and one that is on top of the events affecting its operation. Planning and scheduling provides the foundation for such a maintenance effort.

When we started New Standard Institute, Michael H. Bos and myself had intended to expand on what we had learned from the best efforts at Stauffer. The first new training we developed was maintenance planning and scheduling. We provided it as a public seminar and then developed industry-specific versions of the course, creating a workshop for planner/schedulers as well as the opportunity to consult directly with us, outside of the plant environment. For the past 15 years, the seminar has been presented to more than 5000 maintenance professionals from all over the world. This book has been a work in progress over that time, building on input from all those we have worked with.

Managing maintenance planning and scheduling will always be a requirement of a maintenance effort. Quality and reliability improvement programs have come and gone, as have computerized maintenance management systems, but the basic concept of maintenance planning and scheduling still remains: *Remove the obstacles from performing work, and a maintenance effort will be successful.*

Chapter 1

Defining the Level of Maintenance

Maintenance of a plant or facility can be performed by default or by plan:

- *Maintenance by default* simply means equipment is repaired as it fails—usually on an emergency basis. The rush to get the equipment running again may result in shoddy workmanship that costs in the long run. The downtime incurred by the operation usually occurs at an inopportune moment and can also cost money because of lost business.

- *Maintenance by plan* simply means that there has been forethought in what level of maintenance is required. The level of maintenance eventually attained by a plant or facility depends on the condition of the equipment and the ability of that equipment to meet the needs.

To attain some control over the level of maintenance, organizations must move away from maintenance by default to maintenance by plan. A comprehensive planning and scheduling effort provides the basis for this control. Companies that have moved to a more planned approach tend to be better equipped to meet the numerous market changes encountered in recent years. However, the implementation of planning and scheduling at plants and facilities has had to compete with numerous other programs in recent years.

Any discussion of planning and scheduling must start with a discussion about these programs and how they support, or contradict, the move toward a planned level of maintenance. This chapter provides a high-level view of how you define, plan, and implement a level of maintenance.

Era of Business Management Theories

The period from the 1960s to the present is often referred to as the era of business management theories. Dozens of programs with three-letter acronyms, began cropping up in the 1960s, including the following:

- MBO—Management by objective
- MBR—Management by results
- TQM—Total quality management

- SPC—Statistical process control
- JIT—Just-in-time
- TPM—Total productive maintenance
- BPR—Business practice re-engineering
- RCM—Reliability-centered maintenance

Management consultants made millions pushing the program-of-the-month, sometimes contradicting the program they pushed the year before. In spite of this self-serving agenda, many companies were still able to make real progress toward improving the way they did business. Of all the efforts implemented during this period, none has been more effective than total quality management (TQM).

Prior to the 1970s most quality-control efforts were centered on testing or inspecting the final product. If a product failed an inspection, it was discarded or reworked. Beginning in the 1970s, Japanese companies implemented what they called total quality management. Central to this effort was the reduction of defects in the process, at every point in the process. In fact, the goal was to have zero defects. Production workers were trained in statistical process control (SPC) and were given conditional autonomy to make changes that ultimately led to eliminating defects.

This new focus on eliminating defects not only succeeded in improving the throughput of manufacturing processes, but also had the side effect of reducing the overall cost of manufacturing. Japanese products became more reliable and cheaper than comparable products throughout the world. As a result, many North American and European companies adapted TQM in the 1980s. These efforts were largely successful and are credited with improving the overall quality of products and services throughout the world.

TQM touched every aspect of business. One TQM offshoot, just-in-time (JIT) inventory management, focused on a goal of zero raw material, work-in-progress, and finished goods inventory. Another offshoot was total productive maintenance (TPM), which extended the idea of autonomy and suggested that production line workers could also perform maintenance on equipment. One goal of TPM was to move preventive maintenance (PM) activities from an organized maintenance department to the production worker. The ultimate (and some may say far-reaching) goal was to move all maintenance activities to production personnel. This approach has had limited success in industry, mostly because of implementation problems. In some cases, production workers weren't properly trained to perform the work. In others, the maintenance departments with poorly established PM programs had nothing of substance to pass on to production workers.

By the 1990s, the zero-defect tenant of TQM started to lose its unconditional supporters. Quality for the sake of quality, with no eye toward the bottom line, was said to be just plain bad business. A new approach called business practice re-engineering (BPR) became the rallying cry for industry.

Based on a book by James Champy and Michael Hammer, *Reengineering the Corporation* (New York: HarperCollins Publishers, PerfectBound electronic books, 2003), BPR seemed to address the deficiencies in TQM by re-engineering a process to eliminate non–value-added activities. This often resulted in streamlining the process, eliminating steps, and eliminating people. The wave of downsizing in the 1990s was blamed, perhaps unfairly, on re-engineering. BPR fostered the development of self-directed teams that signaled the end of some first-line and middle-management jobs.

The Movement to Reliability and Availability

A positive outgrowth of TQM and BPR was a movement toward improved equipment reliability. Many organizations began to develop programs, such as reliability-centered maintenance (RCM), to refocus maintenance on overall equipment reliability. RCM was developed in the electric power industry and has been adapted to a number of other industries. RCM founded a renewed emphasis on preventive maintenance (PM) programs and began to give credibility back to these efforts. Originally developed in the 1970s, predictive maintenance (PDM) programs were beefed up or expanded. Standards for purchase, installation, and repair of equipment were developed to ensure continuity as organizations continued to evolve.

However, just as quality for the sake of quality is unprofitable, reliability for the sake of reliability is equally wrong. Many maintenance reliability programs tend to take a bottom-up approach. Reliability tools are purchased and then are applied to any and all equipment to improve performance. A program such as this can result in costly efforts on some less-important equipment. The cost associated with improving reliability must be balanced against the return from the effort.

In a smart business, the market is thoroughly researched and a comprehensive sales plan is developed. An operations department uses this sales plan to determine what equipment availability is required. A maintenance department is consulted to determine what equipment reliability is required to provide the needed equipment availability. The following example illustrates this approach.

Once market share is determined, a sales forecast is made. Estimating the seasonal changes in sales and how they relate to plant capacity helps provide a production forecast. The plant capacity is the nameplate or design capacity, running 24 hours per day, seven days a week. Figure 1-1 shows a sales forecast for a year. (Each month has been assigned an equal number of hours to simplify the example.)

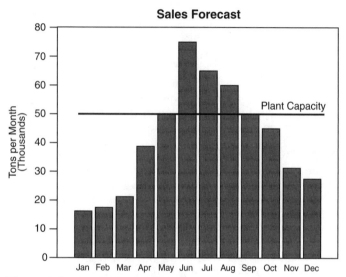

Figure I-I A sales forecast for I year.

The first step in developing a production plan is to determine if enough plant capacity exists to meet the sales forecast. During the months of June through August, sales requirements exceed plant capacity. Therefore, this company must develop a production plan to build inventory and eventually meet sales requirements, even during these months.

The plant capacity line shown on the chart suggests the plant can produce at this level all the time. However, the nameplate rating on any equipment (or plant for that matter) comes with an unwritten caveat. The plant will run at capacity, but not indefinitely. Equipment will have to be shut down for maintenance from time to time. So, to achieve the operations availability, maintenance must also get custody of the plant equipment. Figure 1-2 shows a production plan with this point in mind.

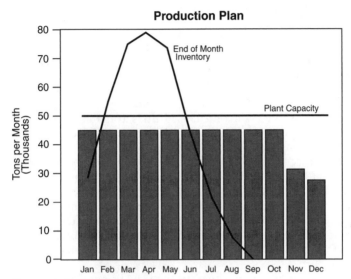

Figure 1-2 Production plan for 1 year.

The plant is scheduled to run below capacity each month, but the production level in the first few months is higher than the sales requirement. This allows for inventory to be built up and eventually meet the sales requirements of the peak months. Operation below capacity during each month allows downtime to be built into this plan. The maintenance department is charged with using this downtime to build capacity back into plant equipment.

With this production plan, the maintenance department can now discuss an intelligent schedule for operations uptime and maintenance downtime, called a *custody plan*. This plan identifies the custody of the equipment required by the operations department to manufacture a quality product and to attain the onstream-time needed to meet sales. It also defines the custody of equipment required by the maintenance department to build capacity back into the operation.

Planned Maintenance under Attack

Maintenance departments in facilities without such a custody plan receive custody of equipment only when it fails—this is maintenance by default. Breakdowns do not necessarily occur at the most opportune moments and tend to result in the most expensive maintenance work performed. Consider the example of a plant that runs around the clock. Studies have shown that only 24 percent of plant breakdowns occur during normal maintenance department work hours (Monday through Friday, 8 hours per day). This is not surprising, since this period accounts for 24 percent of the time in a normal workweek. For this reason, it should not be considered coincidental that the majority of breakdowns (76 percent) occur in the evening or on weekends, when less maintenance help is available. Usually, repairs performed by the maintenance department during breakdowns fall short of being planned, to say the least.

Maintenance by default, however, creates a plant environment that ultimately can be the demise of the business. When the maintenance department receives custody of the machinery only when it breaks down, its marching orders are to get the plant running again. This is a far different mandate from "fix the real problem." To get the plant running again, the repairs that are made are almost always repairs that treat the symptoms of the problems rather than the real problems.

A typical example of this situation would be a plant that originally was designed for a given production capacity, with all motors sized to operate within their nameplate capacity. Increases in production capacity with subsequent debottlenecking of the plant now have many motors running in an overload condition. The effect of this demand is a drastic reduction in motor life, sometimes burning up in a matter of months.

Under maintenance by default, the easiest and quickest fix is to replace the motors in kind, but the replacement motors will fail just as quickly as their predecessors did. The real fix, in this example, would require the next-larger horsepower motor, but this might require modification of the base to accept a larger frame, running new conduit, and/or pulling larger wire. It may also require a larger motor starter in the motor control center. Also, in maintenance by default, there is never time to fix the real problem. In a short while, other equipment problems will be causing additional emergency downtime. Eventually, the entire maintenance department may be consumed in applying bandages to many problems, but not really fixing any.

Some companies deal with this unfortunate scenario by equipping the maintenance department to be more responsive to emergencies. Outfitting the workers with radios or pagers enables maintenance supervisors to locate their personnel at a moment's notice so that personnel can be dispatched at will. Other companies supply computer terminals at operating machinery sites and install printers at maintenance worker benches so that machine operators themselves can send a request directly to a maintenance worker, bypassing any management control altogether. Some companies outfit maintenance shops with answering machines or voice-mail–boxes. Maintenance workers are required to check in regularly for messages and respond immediately.

All of these creative ideas ensure that emergency response is optimized, but do nothing to improve operating reliability. In fact, dealing with emergencies in this way only exacerbates problems and perpetuates emergency thinking. When maintenance focuses only on providing a quick response to emergency conditions, it ultimately removes the need to identify beginning equipment problems as anyone's responsibility. Short-term thinking begins to be the norm and equipment reliability degrades.

If a custody plan exists, maintenance knows when the equipment will be made available and has time to plan the jobs to ensure the highest quality repair. Using its custody of the equipment wisely becomes the challenge of a well-managed maintenance department. This sort of effort tends to separate the work performed into two broad categories:

- Maintenance work performed during maintenance custody (shutdowns)
- Maintenance work performed during production custody

Maintenance Work Performed during Maintenance Custody (Shutdowns)

The following defines maintenance work performed during maintenance custody (that is, during a shutdown):

- *Preventive maintenance (PM)*—The PM work performed during a shutdown of equipment takes the form of manufacturer-recommended rebuilds, or rebuilds required because of predictable wear. Compressor and pump rebuilds, along with instrument calibration, are in this category. The *PM interval* (the time between PM shutdowns) is established from a manufacturer's requirements or a plant operating history. The job can be planned far in advance of the shutdown and should be the most efficient effort by the maintenance department.

- *Corrective work*—This is the work performed to replace worn parts, adjust loose equipment, prevent a major failure, and return the equipment to nameplate condition. As with PM, this work can be planned in advance of the shutdown. To ensure that repairs are performed to at least a minimum standard, all replacement parts, tools, and sufficient labor required for the repair must be allocated prior to the shutdown.

- *Emergency repair*—This is the occasional and unavoidable shutdown of equipment because of unforeseen circumstances requiring unplanned repairs. However, the fact that these repairs were unforeseen may spur maintenance-engineering efforts to add preventive or predictive checks, hence reducing future breakdowns.

Maintenance Performed during Production Custody

The other work that the maintenance department does while production has custody falls into the following broad categories:

- *Prefabrication (Prefab)*—Prefab work includes (but is not limited to) rebuilding equipment in the shop, laying out pipe for future replacement, and setting up equipment prior to a planned shutdown.
- *Preventive maintenance (PM)*—Adjustments, lubrication, tests, inspections, and calibrations performed on equipment while it's running are in this category. PM work is performed on a periodic basis. The optimum PM interval is determined by the maximum period of time that can elapse before failure occurs.
- *Predictive maintenance (PDM)*—This includes nondestructive tests performed on the equipment to determine its condition. By comparing the test measurements to established engineering limits, the need for corrective work can be determined. The limits are set to ensure sufficient time is available for repair and to prevent an emergency shutdown of the equipment.
- *Corrective work*—Plannable repairs on installed spares fall into this category.
- *General maintenance*—This is the non–production-related work (such as work performed on real estate, lawns, roofs, and so on).

NOTE

Preventive maintenance (PM) and predictive maintenance (PDM) programs are discussed in detail later in Chapter 4.

Implementing a custody plan may require approval from upper management, who will want to see the financial justifications and reasoning behind this new approach. Luckily, a custody plan lends itself to a more proactive maintenance budgeting process with justifiable line items.

The Maintenance Budget

A maintenance department's budget is not unlimited. The size of the budget must be weighed against the cost of raw materials, production, and overhead. An old accounting term (still used by some companies) for the money spent on maintenance is burden, which insinuates that the maintenance department is a burden on the operation of the facility. In the traditional sense of the word burden, no attempt has been made to estimate what the cost of maintenance would be; it is just paid for as it happens.

Today, most managers realize the importance of the maintenance department in ensuring plant safety, product quality, and the ability to meet sales requirements. The development of a plan that helps maintain the production schedule becomes the joint responsibility of the maintenance and operations departments.

Establishing the maintenance budget is often the first attempt that some facilities make to determine a plan. Although this is not advised prior to developing a custody plan, it is at least an attempt.

Using a custody plan provides major advantages in estimating true shutdown costs. Adding in the nonshutdown work (such as prefab, PM, and PDM) will develop a complete maintenance budget. Figure 1-3 shows an example of such a budget based on the custody plan and parameter definitions discussed earlier.

As long as the total shutdown labor hours (9350 hours in the example) do not exceed the total custody time allowed for maintenance work, all necessary shutdown work can be completed. Add to this number the total nonshutdown labor hours (47,850 hours in the example), and you can determine an estimate of the required work force.

Setting Goals Based on the Budget

In the example budget, the goals defined by the labor hours in each budget category are summarized in Table 1-1.

It is not the intention of this budgeting method to identify all the budget responsibilities a typical manager may have (such as supervision, clerical, perishable tooling, and new equipment). This method does, however, provide controllable guidelines in determining the level of maintenance a facility wants to attain.

	Labor Hr.	Material Cost	Other Costs	Totals
Custody Work (Shutdown)				
Preventive Maintenance Work	4840	$ 100M	$120M	
Corrective Work	2290	$ 40M	$110M	
Emergencies (from History)	2220	$ 40M	$110M	
Total Shutdown (Labor x Labor Rate)	9350 Hours x $25/Hr.			
Total Cost – Shutdown	$ 234M	$ 180M	$340M	$754M
Non-Custody Work (No Shutdown)				
Prefab Work (for Shutdowns):	7700	$ 206M	$ 47M	
PM Work (Defined Inspections)	12,100	$ 558M	$101M	
Corrective Work	14,600	$ 396M	$ 78M	
Predictive Maintenance:	6900	$ 74M	$ 14M	
General Maintenance:	6550	$ 16M	$ 6M	
Total Non-Shutdown (Labor x Labor Rate)	47,850 Hours x $25/Hr.			
Total Cost – Non-Shutdown	$1196M	$1250M	$246M	$2692M
Total Cost	$1430M	$1430M	$586M	$3446M

Figure 1-3   Budgeting worksheet.

Table 1-1   Goals

	Hours	*Percent*
Emergency	2220	4
Corrective (shutdown and nonshutdown)	16,890	30
PM and PDM (shutdown and nonshutdown)	23,840	42
Prefab work	7700	13
General maintenance	6550	11
Total	57,200	100%

The Purpose of a Work Order System

The maintenance department must be bound by a set of rules and controls that limit the work performed to that which has short-term, near-term, and long-term benefit for the company as a whole. The *work order* is the center of this control. Imagine a facility without a formal work order. The maintenance department would most likely carry out maintenance work based on verbal requests, requests jotted on pieces of paper, or sent via email. Based on the importance of the job requested, the work is either assigned for immediate execution, or could be put on a list for later action. Whenever a maintenance worker becomes available, the next most-important job on the list is assigned. The list of work will tend to grow as more and more requests are made. Some operations personnel may even bypass the list and grab the next available maintenance worker to perform a job that just came up.

Under this arrangement, no attempt is made to quantify the costs associated with the work. Perhaps one piece of equipment is less reliable than it should be, but there would be no data to back that assumption. Additionally, some verbal requests may not even be valid or may be inconsistent with company goals. Requests for expensive modifications or replacements continue on to completion without any review of the costs versus benefit. A verbal request to correct a problem may get lost in the confusion and never make it to the list. The problem may turn into a failure, or could lead to an unsafe condition.

A formal work order system answers these deficiencies and more by providing financial structure to the work that the maintenance department performs. A work order system serves the following purposes:

- Provides a formal means of requesting maintenance services.
- Provides an authorization process that verifies the need or benefit of the service request.
- Provides a vehicle to prioritize and schedule maintenance work.
- Provides the starting point when planning work.
- Aids in collecting labor, material, tool, and special equipment data, as well as costs by cost center, department, or equipment.
- Tracks maintenance work from initiation to completion.
- Provides a coherent written document for the maintenance worker who performs the work.

Objections to a Formal Work Order System

Many maintenance supervisors and team leaders complain they have too much paperwork, too many meetings, and not enough time to visit job sites. They may feel the addition of a work order system will just add more record-keeping duties to their workloads.

Some maintenance workers may resist the implementation of a formal work order system for fear that it could be used as a club against them. For some, these fears have proven to be real. Their managers reframe the central purpose of a work order system as a labor performance-measuring device. Out of all the potential uses of a work order system, this has proven to be the most destructive.

The data provided by a work order system is not performance data. Using a work order system as a performance-measurement tool is destructive in many ways. Most managers would like the support of their employees when collecting historical records or cost data. However, employees are not easily inclined to contribute to a system that can ultimately lead to their terminations.

Work Order Categories

Work order data should be limited to only that which is essential. Information collected on a work order can be categorized three different ways:

- Request and authorization
- Planning and scheduling
- Execution and closure

The following data falls into the request and authorization category:

- Work order number
- Originator/phone
- Approvals
- Date initiated
- Downtime requirement
- Date/time/shift available
- Priority
- Class of work
- Required completion date
- Equipment number and name
- Location, area, building, department
- Account number or cost center
- Work requested (description of the problem)

The following data falls into the planning and scheduling category:

- Planner's name
- Work to be performed
- Job steps (an additional planning sheet may be needed)
- Craft required
- Sequence for this craft (optional)
- Estimated hours
- Parts and material requirements
- Special equipment needed
- Special tools needed
- Contractor requirements (include a cost estimate)
- Permit requirements
- Safety requirements
- Drawing and documents (listed or attached)
- Estimated labor costs
- Estimated material costs
- Estimated contractor costs
- Estimated rental costs

The following data falls into the execution and closure category:

- Supervisor's name
- Comments (by the worker or supervisor)
- Actual hours required for completion
- Failure codes (required by some computer systems)
- Parts and material used (including cost)
- Actual contractor costs

The work order in Figure 1-4 incorporates many of the items described here.

The top half of the example work order is the request and authorization category. The shaded areas must be filled out by the originator. Unshaded areas are optional, but could be filled out to expedite the planning and scheduling phase. The originator of this work order has indicated that the downtime for this job could be made available on 7/25/97 at 8:00 A.M.

The next part is the planning and scheduling requirements. The planner has developed a rudimentary plan for the job. The labor hours required to perform this job have been estimated, as well as

Work Order

W/O # 07344

Date Initiated	Originator	Approvals	Priority	Date Req.	Downtime Requirement	Availability		
						Date	Time	Shift
7/7/97	T. Orkin		R3	7/25/97	Plant 2 S/D	7/25/97	08:00	1st

Equip. #	Equipment Name		Location	Account or CC	Class
25P203	Main Process Pump		Plant 2	CC 25	Corrective

Work Requested (describe the problem)

Pump no longer achieves the pressure it used to, lowering the sprayer efficiency.
Troubleshoot and repair the problem.

Planner	Supervisor	Est. Hours	Craft(s)
Avondale	Layton	32	Mech.

Work to be Performed	
After a dead head test, it was determined that the pump impeller needs to be replaced. Remove and rebuild pump during next Plant 2 shutdown. Also replace guard.	Labor Costs (Est.) **$1000**
	Mat. Costs (Est.) 350
	Contractor Costs (Est.) 0
	Other Costs (Est.) 450
	Total Est. **$1800**

Special Equipment   Crane (Rental)

Parts and Material
Bearing and seal in storeroom. Rebuild kit in planner's office. Purchased sheet steel for new guard. To be delivered 7/8/97.

Contractor Requirements

Special Tools Needed

Drawing and Documents Required

Permit Requirements	Safety Requirements
Line entry, lock-out/tag-out.	Contact operator prior to start.

Comments	Actual Hours
Removed pump, rebuilt it. Impeller was very eroded. Installed new impeller. Coupling had to be replaced. Also fabricated and replaced the guard.	32
	Closeout Date
	8/30/97

Figure 1-4 Work order.

an estimate for the cost of labor, materials, and crane rental. The planner has also identified where the parts can be found, as well as some permit and safety requirements.

The execution and closure portion of the work order is at the bottom. Here, the mechanics have identified some of the work they performed. The actual hours spent working have been entered by their supervisor, along with the date the work order was eventually closed.

The work order should be the basis for a total maintenance information system, which includes the purchasing, inventory management, and payroll systems. Figure 1-5 illustrates how the work order connects to other records and forms in the maintenance information system.

Generating a Work Order

There are a few ways a work order is generated. Larger organizations require that a work request be generated before any work order can be opened. In smaller organizations, the work order doubles as a work request.

NOTE

In the case of emergency work, a written work request is sometimes provided after work begins.

Usually, anyone is allowed to write a work request, but these requests are passed through an approval stage to validate the need or to cull out duplicates. Once approved, the work request can go directly to the work order stage if no plan is required, or if planning is just not practical (as is the case with an emergency). A plan, which includes a filled-in planning sheet, may be required as an intermediate stage to ensure that parts and other resources are available when it comes time to perform the work. A dollar estimate for the requested work should be developed and approved prior to moving the request to the work order stage.

A work order can also be generated from a PM system. A PM schedule is needed to identify the equipment requiring PM, the work or inspection to be performed, and the frequency of the activity. Once the PM comes due to be performed, a related PM task description is added and a work order is generated. Once the equipment, frequency, and task information are stored in a database, the generation of PM work orders is automatic when employing computerized maintenance management systems (CMMSs).

NOTE

Computerized maintenance management systems (CMMSs) are discussed in more detail in Chapter 5.

Many companies require that a purchase request be filled out prior to the generation of a purchase order. Purchase requests go through an approval process to reduce the flow of paper work to a purchasing agent and can facilitate some control over maintenance

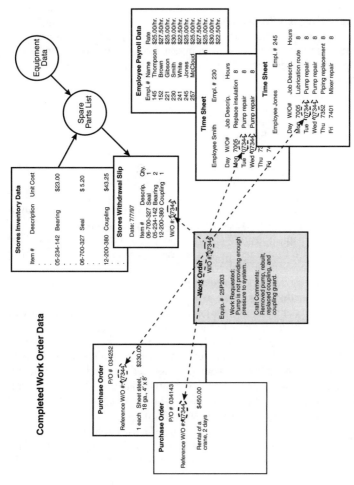

Figure 1-5 Work order connections to other records.

costs. Individuals are provided with a grant of authority, which allows them to commit budgeted funds to purchases. If an individual does not have a high enough grant of authority, the request must be circulated to individuals with consecutively higher grants of authority.

Purchase orders or contracts for services are legal documents between the company and the outside world. A purchasing agent collects purchase requests and carries out the purchase. Technically, anyone in the company can be a purchasing agent. The law of agency states that anyone who makes a purchase for which the bill is paid is a bona fide purchasing agent.

Cost Data

As stated previously, the work order is the focus of most maintenance record-keeping. As such, a work order number should be referenced on every purchase request, stores withdrawal slip (indicating inventory removal), or contract. This allows costs to be tracked and accumulated against a work order. The timesheet is used to collect payroll data. Sometimes the daily schedule doubles as a timesheet.

A connection can always be made between plant equipment and the cost to maintain it, as long as an equipment number or accounting cost center is added to each work order. This information can be used later to determine the total cost of maintaining each piece of equipment.

The system described previously can be completely manual or can be completely paperless. Prior to the widespread use of CMMSs, large banks of filing cabinets were required to store work orders and associated paperwork. A copy of a completed work order was filed numerically, another chronologically, and yet another copy by equipment number. A work order filed by equipment number usually had a copy of all associated timesheets, stores withdrawal slips, and purchase orders filed with it.

Anyone who wanted to know how much was being spent on a certain piece of equipment had to first look up the equipment in the equipment file. Next, the actual cost data for each document filed under that number had to be researched in the accounts paid files and in the payroll records. Once this was done, the cost data could be tabulated to present the actual costs accumulated. Needless to say, this task was not performed often and most likely was passed down to a clerk-level employee.

Getting specific equipment-cost data becomes much easier with the use of a CMMS. The work order number can be referenced on all associated records. The computer system will keep track of the connection between these records, as shown in Figure 1-6.

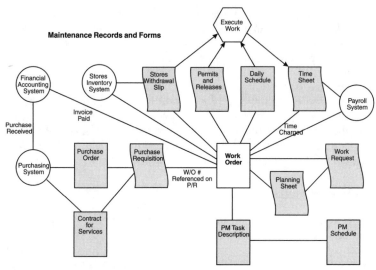

Figure I-6   Work order connections via a CMMS.

Consider a sample scenario. A work order (for example, number 07344) is generated for work performed on a pump with an assigned equipment number (for example, 25P203). Five additional records are associated with the work order. Two maintenance employees worked on this job for two days. Their timesheets show the hours they worked, and their employee numbers reference these records to the hourly wage data in the payroll department. Two purchase orders are also generated as a result of this work order. One is for some sheet steel and the other was for a crane rental. The maintenance workers generate a stores withdrawal slip to remove three items from stores. The item numbers for these parts are referenced back to the storeroom inventory data record, which has the unit cost for each item. Figure 1-7 shows the result of a typical report that a manager might request for this sample work order.

Most CMMSs keep track of the relationship between work orders and associated records as shown in Figure 1-6. The current cost of labor, parts, materials, contracts, and rentals at the time of the closure are also stored when the work order is closed. Reports can then be easily generated for any stored piece of information. Figure 1-8 shows an example of a year-to-date report for the sample equipment number 25P203.

Work Order #: 07344
Labor

7/7/97 Emp. # 230	8 hrs @ $30.00/hr.	$ 230.00
7/7/97 Emp. # 245	8 hrs @ $25.00/hr.	$ 200.00
7/8/97 Emp. # 230	8 hrs @ $30.00/hr.	$ 230.00
7/8/97 Emp. # 245	8 hrs @ $25.00/hr.	$ 200.00

Total Labor $ 860.00

Material - Stores

7/7/97	05-234-142 Bearing	2 ea @ $23.00 ea	$ 46.00
7/7/97	06-700-327 Seal	I ea @ $ 5.20 ea	$ 5.20
7/7/97	12-200-380 Coupling	I ea @ $43.25 ea	$ 43.25

Material - Purchased

8/4/97 (date paid)	P/O # 034252 Sheet Steel	$ 230.00

Total Material $ 324.45

Contractors and Rentals

8/8/97 (date paid)	P/O # 034143 Crane Rental	$ 450.00

Total Contractors and Rentals $ 450.00

Work Order #: 07344 Total $1634.45

Figure 1-7   Typical report for a work order.

The year-to-date total maintenance expenditure for this piece of equipment is $5,993.18. This kind of information is helpful to a manager in the following ways:

- The manager may conclude that more PM is required on this equipment.
- The cost information can help in determining whether this piece of equipment should be replaced rather than repeatedly repaired in the future.
- The repeated repairs on this equipment (the motor was changed twice and fuses were changed numerous times) may

Maintenance Costs for Equip. #25P203 - YTD

Date	W/O #				Cost	Totals
Labor						
2/9/97	06960	Emp. # 152	8 hrs @ $27.50/hr.		$220.00	
2/23/97	07019	Emp. # 230	8 hrs @ $30.00/hr.		$240.00	
2/25/97	07008	Emp. # 241	8 hrs @ $22.50/hr.		$180.00	
3/9/97	07040	Emp. # 145	8 hrs @ $25.00/hr.		$200.00	
3/19/97	07074	Emp. # 145	8 hrs @ $25.00/hr.		$200.00	
3/20/97	07074	Emp. # 245	8 hrs @ $25.00/hr.		$200.00	
4/20/97	07166	Emp. # 221	4 hrs @ $25.00/hr.		$100.00	
4/20/97	07166	Emp. # 241	4 hrs @ $22.50/hr.		$ 90.00	
4/21/97	07166	Emp. # 221	4 hrs @ $25.00/hr.		$100.00	
5/24/97	07270	Emp. # 152	4 hrs @ $27.50/hr.		$110.00	
5/24/97	07270	Emp. # 221	2 hrs @ $25.00/hr.		$ 50.00	
6/11/97	07320	Emp. # 145	2 hrs @ $25.00/hr.		$ 50.00	
6/11/97	07320	Emp. # 230	2 hrs @ $30.00/hr.		$ 60.00	
6/11/97	07320	Emp. # 241	2 hrs @ $22.50/hr.		$ 45.00	
7/7/97	07344	Emp. # 245	8 hrs @ $25.00/hr.		$200.00	
7/7/97	07344	Emp. # 230	8 hrs @ $30.00/hr.		$240.00	
7/8/97	07344	Emp. # 245	8 hrs @ $25.00/hr.		$200.00	
7/8/97	07344	Emp. # 230	8 hrs @ $30.00/hr.		$240.00	

Total Labor $2,725.00

Material - Stores

Date	W/O #					Cost
2/9/97	06960	12-200-380	Coupling	1 ea. @ $43.25 ea.	$ 43.25	
2/9/97	06960	66-382-810	Motor	1 ea. @ $210.34 ea.	$210.34	
2/25/97	07019	62-110-012	Circuit Break	1 ea. @ $107.65 ea.	$107.65	
2/25/97	07008	65-314-036	Fuse	3 ea. @ $12.65 ea.	$ 37.95	
3/9/97	07040	56-041-250	Valve	2 ea. @ $52.86 ea.	$105.72	
3/20/97	07074	06-700-327	Seal	1 ea. @ $5.20 ea.	$ 5.20	
3/20/97	07074	43-422-896	Sheave	1 ea. @ $40.59 ea.	$ 40.59	
4/20/97	07166	06-700-327	Seal	1 ea. @ $5.20 ea.	$ 5.20	
4/20/97	07166	43-422-896	Sheave	1 ea. @ $40.59 ea.	$ 40.59	
5/25/97	07270	06-700-327	Seal	2 ea. @ $5.20 ea.	$ 10.40	
6/11/97	07320	66-382-810	Motor	1 ea. @ $210.34 ea.	$210.34	
7/7/97	07344	05-234-142	Bearing	2 ea. @ $23.00 ea.	$ 46.00	
7/7/97	07344	06-700-327	Seal	1 ea. @ $5.20 ea.	$ 5.20	
7/7/97	07344	12-200-380	Coupling	1 ea. @ $43.25 ea.	$ 43.25	

Material - Purchased

Date	W/O #			Cost
4/11/97	07008	P/O 033807	Cable, 500'	$176.50
5/9/97	07166	P/O 033965	Local Disconnect	$410.00
8/4/97	07344	P/O 034252	Sheet Steel	$230.00

Total Material $1,728.18

Constractors and Rentals

Date	W/O #			Cost
2/28/97	06960	P/O 033759	Millwright Contractor	$320.00
4/7/97	07074	P/O 033873	Crane Rental	$450.00
7/18/97	07320	P/O 034119	Millwright Contractor	$320.00
8/8/97	07344	P/O 034143	Crane Rental	$450.00

Total Contractors and Rentals $1,540.00
Equip. #: 25P203 Total $5,993.18

Figure 1-8 Year-to-date report for equipment #25P203.

be indicative of the quality of the repair, or the suitability of the equipment for the application.

With a work order and cost management system in place, a workable planning and scheduling effort is the next logical step in a good maintenance program.

Planning and Scheduling

Unfortunately, many maintenance employees feel as though their jobs are filled with interruptions. The most common complaint of maintenance workers is, "They're always pulling me off one job to start another." The second most common complaint is, "I'm often assigned to work on a job that someone else has started." These complaints do not bear out a lower work ethic, as claimed in a number of recent editorials. In actuality, they point to an unfulfilled willingness to start a job and finish it right the first time.

Good planning and scheduling can provide the environment to motivate employees. A job that has been planned correctly provides the parts and equipment required for the job, so the job won't have to be stopped. A job that is scheduled and arranged with operations should start on time and should not make the maintenance worker wait. If more planned jobs are completed on time, the operations department will be more inclined to write work requests far in advance of the needed completion date. This should reduce the number of so called higher-priority jobs that cut into the scheduled jobs. Everyone in the organization can reap the benefits of proper planning and scheduling.

The Planning and Scheduling Profession

Many companies leave the job of planning and scheduling to the first-line maintenance supervisor. At first glance, this is a logical approach. All the work requests may come to the supervisor, so it is assumed that the supervisor is the most knowledgeable on what is required to complete the job and who can best do the work. A supervisor usually can develop some semblance of a plan without much information. As a matter of fact, formal work requests may not even be needed if the supervisor does the planning. Companies without planners usually consider the net result of any plan to be negligible.

Why is it that many companies feel an increasing need to add a dedicated planning function to the maintenance staff? In some cases, it is felt that the jobs can proceed much more efficiently if the work is better supervised. Freeing up the supervisor to supervise rather than plan and schedule work can lead to a 10- to 15-percent reduction in job duration. Just a 5 percent improvement in time on jobs will more than pay the salary of a good planner, who can plan for 20 or more employees.

Job quality may also improve with the addition of a planning and scheduling function. Identifying, purchasing, and allocating parts, materials, and support equipment can limit travel time by maintenance employees. Scheduling and coordinating jobs with the operations department can cut down on waiting time and other operations-related delays. Step-by-step plans on jobs that require operations downtime can reduce the elapsed time equipment is out of service.

Supervisors are often too preoccupied with day-to-day work and performance problems to think of the future. Planners are usually freed up to address the long-term planning needs of a facility. Outages, shutdowns, and turnarounds can be (and should be) planned and scheduled months in advance if a planner is available to think about the work involved. Dedicated planners/schedulers can take the necessary time and save the company from extended downtime.

The position of the planner/scheduler is often filled by people who have a maintenance background. In some facilities, the job is a promotion from the hourly ranks. In others, the planner/scheduler position is a promotion from first-line maintenance supervisor. More often than not, the planner position is on an even standing with a maintenance supervisor.

Some companies split the job of planner/scheduler. This is very common in nuclear utilities in which special requirements of permitting and coordination with operations are divided between the planner and the scheduler. Other companies create a separate scheduler position as a promotion for clerical personnel. In these companies, the planner builds a plan for a job and develops a list of equipment required to complete it. The scheduler uses the labor hours estimated by the planner and the resulting elapsed time required to develop a schedule based on the work force available. Equipment downtime requests of operations and coordination of heavy maintenance equipment are the responsibility of the scheduler.

Other facilities free the planner/scheduler from the requirement to purchase and expedite parts and materials. A formal expeditor position is developed to meet this need. All the planner has to do is identify the parts required; the expeditor does the rest.

No real plan can be developed for emergency work. A company that plans emergencies is usually planning on emergencies. For this reason, most companies assign the responsibility to handle emergency work to the first-line supervisor. The planner may help the supervisor from time to time, but should never be assigned to handle emergencies. Planners/schedulers who have emergency dispatch as part of their job descriptions usually find they have time for little else.

A planner/scheduler's time is usually broken down as follows:

- *Planning and estimating jobs*: 20–30 percent
- *Scheduling*: 15–25 percent
- *Purchasing and expediting part*: 20–30 percent
- *Other duties*: 25–35 percent

Planning and estimating maintenance work should be the largest part of the planners' day. Scheduling usually involves phone discussions with operations or meetings with both operations and maintenance supervisors. Unfortunately, purchasing and expediting parts can take up a large part of a planner/scheduler's time. Some parts may not be held in the storeroom, so the planner/scheduler must purchase them. Some planners are also required to coordinate contractors or handle a small maintenance crew. This can cut even further into their time to plan jobs.

Planning Defined

True planning of maintenance work is a very distinct effort. It can best be described as follows: *Planning* is the allocation of needed resources, and the sequence in which they are needed, to allow an essential activity to be performed in the shortest time or at the least cost.

Planning requires the identifying and, if necessary, allocating of any resource that is needed to get a job done. Also, the necessary order (or sequence) is needed in the planning of a job even though this information is not really pertinent until the job is scheduled for execution.

The final goal of a plan is to limit cost or time. A plan may be designed to reduce labor hours required to perform a job, thereby limiting the cost of performing the job. If equipment downtime is required for a job, the cost of downtime can be reduced with a good plan. Planning jobs that require downtime may involve balancing the cost of equipment downtime against the cost of additional labor to perform a job more quickly.

Resources

Real planning of maintenance work requires the identification of all the needed resources to complete a job. Until this has been done, there can be no assurance that the job can proceed without delay.

A resource can be anything that is consumed when performing work. Resources to be considered are:

- Labor
- Materials, parts, supplies, and tools
- Support equipment (cranes, lift trucks, and so on)
- Contracted (outside) services
- Time

Labor

Some planners only consider labor hours when planning work. However, real planning requires that crew size also be included. Some jobs may require special skills that preclude just anyone from performing the work. As a result, the allocation of labor may also require identifying exactly who will do the job.

Labor can be identified in months, weeks, or days, but is usually reported in hours. Some estimating methods provide for 0.00001-hour intervals for labor estimates but generally estimates are made at a minimum interval of 0.25 hours.

The cost of labor must also be factored into jobs. This cost may vary, depending on the trade or skill level required, or it can be reported at what is called a *standard cost*, which usually includes benefits, overtime, and indirect overhead.

Materials, Parts, Supplies, and Tools

Identifying the needed material, parts, supplies, and tools that will be used or consumed is just as important as allocating the hours of skilled labor to complete a job. Implied in the planning process is verifying that needed items are, in fact, available and dedicated to the job in question.

Material and parts have a cost that should be identified in the plan. The full magnitude and effect of a job on the maintenance budget is not known until both labor and material costs are added.

Tools identified can be limited to special tools (such as torque wrenches or sledgehammers) not normally carried in a maintenance worker's toolbox. Other planners may feel the need to identify every tool, down to the size and type of screwdriver.

Support Equipment

Identifying and allocating support equipment is often not considered in the planning of maintenance work. The result of this oversight is usually workers waiting until the necessary support equipment becomes available.

Equipment such as fork trucks, pickup trucks, or cranes can be in limited supply. Identifying their need in a job is the first step to ensuring availability when the job starts. If the equipment is rented, the rental cost should be identified in the plan. If the equipment requires a trained operator, this individual's labor hours and cost must be included in the plan.

Contracted (Outside) Services

Contracted labor may be required from time to time to supplement the existing work force. For example, a large pipefitting job may exceed the available workforce's ability to complete the work within a reasonable period of time. A facility may elect to work overtime to reduce this backlog of work, but the cost of overtime may exceed the cost of hiring an outside pipefitting firm to complete the job.

Contracted labor may also be used to supplement facility personnel during shutdowns, outages, or turnarounds. Staffing up using contract labor can ensure that these events are completed on time.

Sometimes the completion of a job requires special services. One typical example includes an annual boiler inspection. At some point in the job, a boiler inspector will have to be called in to make the inspection to satisfy the regulatory agencies. The job cannot proceed as planned and downtime will be extended if this resource is not allocated.

Time

One resource that must be allocated is the elapsed time that should be available to perform the necessary work. Many labor estimates include only the total labor hours for the job. If more than one person is assigned to complete the job, the elapsed time should be less than the total estimated labor hours.

Operations must be advised of the elapsed time required for a job, especially if the job requires that equipment be shut down.

Scheduling Defined

Scheduling is also a distinct function from planning, but is closely tied to planning. It can be defined as follows:

> *Scheduling* is the assignment of many planned jobs into a defined period of time in order to optimize the use of the resources within their constraints.

Effective scheduling cannot be accomplished without planning. A schedule is just a list of work if the jobs on the schedule are not planned. Implied in the act of scheduling is a method of determining the importance of one job over another. This is called a *priority system*. Any good priority system provides a method of determining which job is the most important, second most important, and so on.

Before discussing priorities, it's a good idea to discuss the constraints on resources that affect scheduling work.

Resource Constraints

Scheduling seeks to optimize the use of resources. For example, jobs can be selected and sequenced to minimize travel time. A number of jobs all requiring the use of a crane, for example, may be slotted throughout the workday in order to utilize the crane most effectively. An effective scheduling discipline also helps determine the trade-off between higher resource costs and extended downtime.

All maintenance resources are constrained to some degree. They are not available in unlimited supply. The following constraints are common:

- Fixed amount of labor
- Limited or special skills
- Space
- Physical properties
- Rules and regulations
- Money

Fixed Amount of Labor

There is an upper limit to the labor hours available at any time. Even though the work force can be augmented with contracted help, this resource may not possess the skills for all work to be performed. Available labor levels actually vary because of vacations, absenteeism, and meetings. Proper short- and long-term scheduling requires that an up-to-date labor calendar be developed.

Limited or Special Skills

Some jobs may require specialized skills. Only a few of the personnel in the maintenance work force may possess the necessary skills. Unavailability of these individuals may prevent the start or completion of some jobs.

Space

This is one constraint that is often overlooked. There is a limit to the number of people who can be working in close proximity to each other. Proper coordination of different crews working in the same area can ease this constraint. In most circumstances, all work performed by one crew must stop before another crew can enter the area.

A unique example of this constraint involves the use of cranes. OSHA law requires that as the lift is being made, no work can occur under the lift path.

Physical Properties

The physical properties of equipment or machinery can also be a constraint. For example, work cannot begin on an internal boiler repair until it has cooled down to a safe temperature. The cool-down period for a boiler is a function of how much heat must be removed and the ability of the boiler to dissipate its heat.

Often, chemical piping and associated equipment must be decontaminated before it can be opened or replaced. The decontamination process may involve pumping a caustic, an acid, or water through the system, or purging with an inert gas. Each of these tasks takes time and can limit the time remaining for repair.

Rules and Regulations

Safety requirements (such as lock-out and tag procedures or confined space permitting) are necessary constraints on the scheduling process. Some jobs require extra personnel (such as a fire watch or flag man) to aid in the safe progress of a job.

Work rules can be a constraint on effective scheduling as well. If strong craft distinctions are in force, coordination of workers with differing skills can extend the job duration.

Money

This is the ultimate constraint. Even when extra help can be contracted to shorten a job's duration, the cost of bringing in this help may be prohibitive. This constraint is determined by weighing the cost of performing the job against the perceived benefit.

Summary

The period from the 1960s to the present is often referred to as the era of business management theories. Of all the efforts implemented during this period, none has been more effective than total quality management (TQM). Beginning in the 1970s, Japanese companies implemented TQM to attain a goal of zero defects in their processes. Production workers were trained in statistical process control (SPC) and were given conditional autonomy to make changes that ultimately led to eliminating defects.

This new focus on eliminating defects not only succeeded in improving the throughput of manufacturing processes, but also had the side effect of reducing the overall cost of manufacturing. Many

North American and European companies adapted TQM in the 1980s. These efforts were largely successful and are credited with improving the overall quality of products and services throughout the world. By the 1990s, the zero-defect tenant of TQM started to give way to a new approach called business practice re-engineering (BPR).

TQM touched every aspect of business, including, just-in-time (JIT) inventory management and total productive maintenance (TPM). A positive outgrowth of TQM and BPR was a movement by many organizations toward reliability-centered maintenance (RCM) to refocus maintenance on overall equipment reliability. RCM founded a renewed emphasis on preventive maintenance (PM) programs and began to give credibility back to these efforts. Originally developed in the 1970s, predictive maintenance (PDM) programs were beefed up or expanded. Standards for purchase, installation, and repair of equipment were developed to ensure continuity as organizations continued to evolve.

The cost associated with improving reliability must be balanced against the return from the effort. An operations department uses a sales plan to determine what equipment availability is required. A maintenance department is consulted to determine what equipment reliability is required to provide the needed equipment availability. Using a production plan, the maintenance department can discuss an intelligent schedule for operations uptime and maintenance downtime, called a custody plan. This plan identifies the custody of the equipment required by the operations department to manufacture a quality product and to attain the on stream-time needed to meet sales. It also defines the custody of equipment required by the maintenance department to build capacity back into the operation.

Maintenance of a plant or facility can be performed by default or by plan: When maintenance by default is in effect, maintenance departments in facilities without a custody plan receive custody of equipment only when it fails. This creates a plant environment that ultimately can be the demise of the business. If a custody plan exists (maintenance by plan), maintenance knows when the equipment will be made available and has time to plan the jobs to ensure the highest-quality repair. Using its custody of the equipment wisely becomes the challenge of a well-managed maintenance department.

A maintenance department's budget is not unlimited. The size of the budget must be weighed against the cost of raw materials, production, and overhead. Establishing the maintenance budget is often the first attempt that some facilities make to determine a plan. Although this is not advised prior to developing a custody plan, it is at least an attempt.

The maintenance department must be bound by a set of rules and controls that limit the work performed to that which has short-term, near-term, and long-term benefit for the company as a whole. The work order is the center of this control. A formal work order system provides a financial structure to the work that the maintenance department performs.

Planning is the allocation of needed resources, and the sequence in which they are needed, to allow an essential activity to be performed in the shortest time or at the least cost. A resource can be anything that is consumed when performing work. Scheduling is the assignment of many planned jobs into a defined period of time in order to optimize the use of the resources within their constraints. Effective scheduling cannot be accomplished without planning.

Chapter 2 builds on the foundation we have built in this discussion and tackles the details of estimating and planning maintenance work.

Chapter 2

Estimating Methods

As discussed in Chapter 1, the planning of a job requires identifying all needed resources to complete the job. Good job plans also provide the framework for proper execution of work and help ensure all resources are used more efficiently. One of the more important aspects of a good job plan is an estimate of the labor resources required.

This chapter discusses the various methods used to estimate time and costs associated with a job. The discussion focuses on several means by which estimating is performed, including construction estimating methods, methods time measurement (MTM), the planning thought process, and using past performance to generate estimates for new work. This chapter also discusses factors that affect the accuracy of an estimate, as well as the parts and material requirements necessary to create a complete plan. To begin this discussion, let's take a general look at the use of estimates in maintenance.

Judgment versus Guessing

All estimating processes have some form of inaccuracy. After all, estimating is an attempt to determine the condition of equipment from minimal data, predict the scope of work or repair needed (which is contingent on knowing the condition of the equipment), and predict the activities of people who are very unpredictable.

Good estimating involves good judgment. An estimate based on judgment is founded on the authoritative opinion of the planner. Estimates based on judgments are more likely to be met than those determined through guesswork. A guess differs from a judgment because an estimate of the time required to complete a job (or job step) is based on little or no evidence. All estimates for maintenance work should be judgment estimates at the very least. The planning process can be improved by improving the planners' judgment.

Many experienced planners refer to an intuition or feeling they have about the time estimates. These are usually not feelings at all but rather a set of rules based on experience that make estimating second nature or automatic. These people are actually performing estimates by judgment.

An inexperienced planner can use the judgment of others to develop a good estimate rather than guessing. Standard estimating books and databases for construction work are available. These references provide some help to maintenance planners, but are seldom the final word. The experience of others in the maintenance department can also be tapped. Other planners, first-line supervisors, and craft employees can be a major help in turning a guess into good judgment.

In addition, the accuracy of an estimate depends on the training and experience of the estimator and the quality of the data available to the planner. Data quality can be improved though investigation of the job and good data-collection techniques.

How Job Estimates Are Used in Maintenance

Planners, supervisors, and managers use estimates developed for maintenance work in various ways. Depending on the form, an estimate can be used to identify and control daily activities of the maintenance work force, or it can be used to control the cost of maintenance in a facility. Estimate types can be divided into two forms: *hour estimates* and *dollar estimates*.

Hour estimates are used in the control of a limited labor resource. The most common reasons for developing hour estimates are:

- *Scheduling*—Daily and weekly schedules are developed based on the direct labor hours assigned to a job and the elapsed time required to complete the job. (Elapsed time may be shorter than direct labor time when more than one person is working on the job.) Elapsed hour estimates may be crucial for operations preparation.

- *Backlog evaluation*—The number of direct labor hours currently identified by work orders on hand are called the *backlog hours*. The backlog hours are compared to the current work force to determine if a sufficient work force exists. Backlogs are often used to adjust the work force up or down, and can also be used to identify the need for overtime and contract work.

- *CPM and PERT data*—Larger jobs or projects usually have time and dollar constraints that must be met. Critical path method (CPM) or program evaluation and review technique (PERT) methods may be employed to improve the coordination of efforts. Accurate estimates of both direct labor time and elapsed time must be determined for each activity on a CPM or PERT schedule.

Hour estimates developed by a planner should never be used to measure worker productivity. Comparisons of actual to estimated hours should be made only to improve estimates. Poor work performance falls directly in the area of supervision and not estimating. Any attempt to match performance to an estimate can lead to resentment, and usually has the opposite effect on productivity than was originally intended.

Dollar estimates are used in the control of the funds available to maintenance. The most common reasons for developing dollar estimates are as follows:

- *Make contract or buy decisions*—Dollar estimates for jobs using facility personnel and materials can be compared to the cost of contracting the work at a lower labor rate. Dollar estimates can also help planners compare the costs for labor and materials to repair equipment to the cost of new equipment. For example, a planner may decide the parts and materials required to rebuild a valve exceed the cost of purchasing a new one.

- *Approvals*—Controls may exist that require an accurate dollar estimate be assigned to a job before work can proceed. The added step of approvals is designed to keep low-priority work (which is not necessarily essential to the facility) from being performed. Persons with higher grants of authority may have to review the job before it can begin.

- *Budgeting*—Long-term planning requires that a budget be laid out for work scheduled for the coming year or years. Dollar estimates on PM and other repetitive work help streamline the budget process. Large shutdowns, turnarounds, and outages are often preplanned, and the dollar estimates for personnel and material requirements are often known.

In many cases, planners are only asked to identify the labor hours, materials, and equipment needed for a job. An estimation of the cost of these items is not part of the estimate at most facilities. The cost of a job is usually important at facilities in which anyone can write a work request. Jobs that seem important to an individual operator or production-line worker may not fit into the goals of the facility as a whole.

Some facilities develop a maintenance budget and then divide the control of the budget among different operating units. The maintenance department is viewed as an outside service organization. This form of budget control moves the ultimate responsibility for maintenance costs away from the maintenance manager and to the operations manager. The operations manager will, in turn, delegate accountability for maintenance costs to the managers (superinten-dents or supervisors) of the different operating units. These de-partment managers may want to know the potential impact a job will have on their budget. A dollar estimate from the maintenance department provides this information.

Other facilities may charge back maintenance costs to the de-partment requesting the work. For example, a maintenance depart-ment may perform work for the engineering department on capital projects that are not considered part of the normal maintenance of the facility. The time and labor for this work is charged to the capital project and not against the maintenance budget. Some facilities go one step further and have the maintenance department bid on capital work. This bid is compared to the bids of outside contractors.

The most common method of controlling the maintenance bud-get and instilling accountability uses a multitier approval system. This system assigns what is called a grant of authority to different individuals in the facility. Each of these individuals is allowed to make purchases or authorize maintenance work only up to their as-signed grant. For example, a planner or supervisor may only have a grant of authority of $1000. Any job or purchase with a dollar estimate over $1000 must go to the maintenance manager for ap-proval. Any job over $5000 may have to go to the plant manager for approval.

Grants of authority are often abused. The tendency to spend what's in the budget is strong, especially near the end of the year. Large jobs or purchases are broken into small parts to avoid going to the next level of approval. This tends to undermine the system and reduces the control on maintenance costs.

Planners often employ several different methods to build a plan for a job. A few of the common ones will now be discussed.

Planning and Estimating Methods

All estimating methods require that a planner be trained in a specific way of thinking. The process should be logical and repeatable. The result of the process should be an estimate that closely approximates the actual steps and time required to complete a job. The following is a list of common estimating methods used in industry today.

- Construction planning and estimating
- Methods time measurement (MTM)
- Planning thought process
- Estimates based on past performance

Construction Planning and Estimating

Those involved in building trades have used structured estimating methods for years. Construction of houses, buildings, and process plants usually require that elaborate estimates be developed and submitted by a number of contractors. The contractor with the lowest bid will get the job.

Some industries (such as pharmaceutical and consumer products) have very active construction programs going on all the time. Planners in these companies are required to estimate the cost of a project and present a plan for its completion. However, a contractor's estimator often has a different goal to achieve from a facility planner.

If the construction estimate is too low when compared to the actual costs, the contractor can lose money. If the job is big and the estimate is too low, the contractor can go bankrupt. If the estimate is too high, the job may go to another contractor. Very often, the lowest bidder loses money on the job. If this happens more than a few times, the construction estimator may be fired. If it happens too often, the contractor may go broke. A facility planner can usually miss a number of estimates without affecting job status.

Additionally, construction estimates should not only include direct costs (labor, materials, and equipment), but also indirect costs. Indirect costs include maintenance of the construction site, insurance, depreciation, legal fees, professional services, and site office expenses. Many of these costs fall under the category of general and administrative (G & A). Overhead can be from as low as 5 percent of the project costs for small firms, to as high as 15 percent for large firms.

Judgment and history are the key tools used in construction estimates. Many resources are available to an estimator if background is lacking in certain areas. The following are common resources used:

* *R. S. Means Company, Inc.*—Construction Consultants and Publishers, 100 Construction Plaza, Kingston, MA, 02364, (800) 334-3509.
* *John S. Page Estimator's Series*—Gulf Publishing Company, Book Division, Box 2608, Houston, TX, 77001, (713) 529-4301.
* *Richardson Engineering Services, Inc.*—P.O. Box 9103, Mesa, AZ, 85214-9103, (602) 497-2062.

These resources provide standard methods for estimating labor, material, and equipment costs. They are restricted to work having to do with demolition or construction. For this reason, they are not helpful when it comes to repair estimates.

An estimator who is unfamiliar with a craft, task, or equipment type can find estimates in these resources. For example, an estimator may have little or no background in excavating methods and equipment. A detailed description of all common bulldozers, graders, and scrapers are listed. Advantages, limitations, fuel consumption, and standard earth-moving rates are listed. Costs of backup equipment (such as water trucks, compactors, and pickup trucks) are also listed, as well as personnel costs for operators and truck drivers. Different sets of estimates are provided for sandy gravel, common earth, or tough clay. The estimator only needs to know how much earth needs to be moved and to where.

Standards are available that provide basic hourly pay rates for various crafts in different geographical locations. This information is important if workers are drawn from a local labor pool. The rates are updated in these publications about every 3 to 6 months.

Estimates for construction work provide one other challenge not required for repair estimates. There is no physical plant to look at while developing an estimate. Construction engineers and estimators must rely on take-offs instead. *Take-offs* are detailed lists compiled from drawings and project specifications, literally taking this data off the drawings. The lists will include all the materials and equipment required for the project. Estimators must train themselves to identify all project requirements from the drawings and site visits.

Two estimators using the same estimating standards may still come up with different estimates. This is possible because they may use different assumptions or methods, and may build a different take-off list. In most cases, however, estimates derived using these standards will be comparable.

Engineering firms hired to investigate and then develop a project usually follow a similar process. The owner or client may require the development of different types of estimates for each step in the life of a project. Following are some common cost estimates required by owners or clients:

- *Budget estimate*—An estimate must be developed to help inform the client if the project is within their budget. The cost of developing such an estimate must be kept low, so the information detail is usually sketchy. A contingency of about 50 percent of the estimate is included in the project because of this lack of data. If the project is accepted, the finished price should not exceed the budget estimate.

- *Predesign estimate*—The design information has not been improved as yet, but a client may want further background information from potential construction firms who may be awarded the project. The contractors will usually put a 50-percent contingency into their estimate.

- *Conceptual estimates*—When design information is available, an engineering firm requests conceptual estimates. An estimator must review the drawings and provide a number that will be used for comparison purposes in the future. Rough tables and charts are used to price equipment and labor, such as the one shown in Figure 2-1.

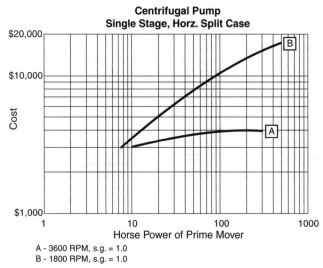

Figure 2-1 Conceptual estimate chart example.

For example, if the job required 12 single-stage centrifugal pumps, the cost for the pumps would be calculated as shown in Table 2-1.

Preliminary Estimate

At this point, the estimator will try to develop detailed estimates for certain aspects of the job to firm up the estimating process (such as labor, materials, and equipment). There may be some items where a good estimate cannot yet be derived. The estimator may identify

Table 2-1   Calculating the Cost of Pump Installation

			From Chart	
Qty.	Size	Speed	Per Unit	Total
5 each	100 HP	3600 RPM	$4000	$20,000
3 each	50 HP	1800 RPM	$8500	$25,500
4 each	25 HP	3600 RPM	$3500	$14,000
			Total Cost	$59,500

A 30-percent contingency is usually included in this estimate.

that part of the estimate as a plug-in or plug number. A *plug* is a temporary figure that is used until a better estimate can be developed. Preliminary estimates usually include a 25-percent contingency.

Design Review Estimate
A *design review estimate* is an extension of a preliminary estimate with as few plugs as possible. The sources of information (such as labor rates and crew size) must be provided. A 20-percent contingency is usually included in the estimate.

Preconstruction Estimate
All plugs must be resolved. The estimator must complete all unestimated parts of the project. A 10-percent contingency is included in the total estimate. Lump sum or unit price costs may be supplied to the client at this time.

Construction Estimate
A *construction estimate* is a complete reassessment of the project prior to construction. All labor, materials, and equipment are identified. Vendor quotes and subcontractor bids are also spelled out. Only a 5-percent contingency is included in this estimate, so it must be as accurate as possible. Construction estimates are used for cost control purposes during construction. Actual costs are compared to the estimate as a job progresses to let the contractor know what parts of the job are profitable and what parts are getting out-of-hand.

Sample Construction Estimates
The following example illustrates a construction estimating method. Assume the isometric drawing Figure 2-2 must be used to shop-fabricate a simple bypass loop.

Overall dimensions are shown rather than the actual pipe and fitting lengths. The take-out dimensions for the fittings must be

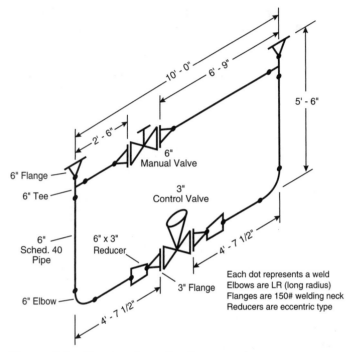

Figure 2-2   Construction estimating example.

subtracted from the overall lengths. Table 2-2 shows the fitting and take-out dimensions used to determine the amount of pipe required.

To determine the time to complete the project and the amount of pipe required, each side should be calculated separately and then totaled. For example, the height of the image measures 5 feet, 6 inches (66 inches). Each side contains a 6-inch diameter pipe, two straight tees with a take-out dimension of $5^{5}/_{8}$ inches, a 150# welding-neck flange with a take-out dimension of $3^{1}/_{2}$ inches and a 90° long radius elbow with a take-out dimension of 9 inches. The result is pipe with a length of $42^{1}/_{4}$ inches (3 feet, $6^{1}/_{4}$ inches). Table 2-3 shows the resulting pipe lengths for all the sections.

A labor estimating table is used to calculate the hours to fabricate the piping loop. The one shown in Table 2-4 provides an estimate for fabricating 150# and Schedule 40 Butt Weld pipe and fittings.

Fitting time estimates are for the total weld time on the fitting, not the time per weld. Care must be taken not to add time when two fittings are welded together, as opposed to a fitting to a piece of pipe.

Table 2-2 Sample Fitting and Take-Out Dimension Table (Butt Weld—150, 300, and 600# Class)

Fitting		Pipe Size—Nominal (Inches)				
		2 Inches	3 Inches	4 Inches	6 Inches	8 Inches
Elbow 90° SR	c-e	2	3	4	6	8
Elbow 90° LR	c-e	3	$4^1/_2$	6	9	12
Elbow 45°	c-e	$1^3/_8$	2	$2^1/_2$	$3^3/_4$	5
Offset (2 45°s)	c-c	$1^{15}/_{16}$	$2^{13}/_{16}$	$3^9/_{16}$	$5^5/_{16}$	$7^1/_{16}$
Tee—Straight	e-e	$4^{11}/_{16}$	$6^{13}/_{16}$	$8^9/_{16}$	$12^{13}/_{16}$	$17^1/_{16}$
Tee—Reducing	c-e	$2^1/_2$	$3^3/_8$	$4^1/_8$	$5^5/_8$	7
Tee—2-Inch Branch	c-e		3	$3^1/_2$		
Tee—3-Inch Branch	c-e			$3^7/_8$	$4^7/_8$	
Tee—4-Inch Branch	c-e				$5^1/_8$	$7^1/_4$
Tee—6-Inch Branch	c-e					$6^5/_8$
Reducer	e-e		$3^1/_2$	4	$5^1/_2$	6
Flange—Weld-neck 150#		$2^1/_2$	$2^3/_4$	3	$3^1/_2$	4
Flange—Weld-neck 300#		$2^3/_4$	$3^1/_8$	$3^3/_8$	$3^7/_8$	$4^3/_8$
Flange—Weld-neck 600#		$3^1/_8$	$3^1/_2$	$4^1/_2$	$4^7/_8$	$5^1/_2$
Flange—Slip-on—All Classes		Wall thickness of pipe + $^1/_{16}$ inch				
Lap Joint—All Classes		6	6	6	8	8

So, only the tee estimate needs to be calculated, because it includes the weld to the 6-inch flange. The same is true for the reducer and the 3-inch flange. The resulting estimate is shown in Table 2-5.

The pipe handling time is a catch-all number, which assumes that the time to cut, bevel, and lay out the pipe in the shop is proportional to 0.59 times the total pipe required (about 21 feet).

Some planners may find an estimate of 52.4 hours on the high side, when compared to their experience at their own facility. This should not deter them from using the published tables, because they can scale the estimate based on actual experience. Assume the actual time charged to the job in the example was only 31.5 hours, or about 60 percent of the original estimate. In the future, the planner can use the estimate developed from the published table, multiplied by 0.6.

Table 2-3   Pipe Lengths

Run	Fitting	Less	Section	Qty	Totals
2 feet, 6 inches	6-inch Tee	5⅝ inches	1 foot, 8⅞ inches	1	1 foot, 8⅞ inches
	6-inch Flange	3½ inches			
6 feet, 9 inches	6-inch Tee	5⅝ inches	5 feet, 11⅞ inches	1	5 feet, 11⅞ inches
	6-inch Flange	3½ inches			
5 feet, 6 inches	6-inch Tee 5⅝ inches × 2	11¼ inches	3 feet, 6¼ inches	2	7 feet, ½ inch
	6-inch Flange	3½ inches			
4 feet, 7½ inches	6-inch Elbow	9 inches	3 feet, 2¼ inches	2	6 feet, 4¼ inches
	6-inch × 3-inch Reducer	5½ inches			
	3-inch Flange	2¾ inches			
				Total	21 feet, 1¾ inches

Table 2-4 Sample Labor Estimating Table (Hours per Unit)

Material Activity	2 Inches	2 ½ Inches	3 Inches	4 Inches	6 Inches	Unit
			Nominal Pipe Size			
Pipe Handling Time Installed on Yoke and Roll						
Hangers—10-foot OC	0.26	0.34	0.37	0.43	0.67	Lineal Foot
Shop Fabrication	0.21	0.27	0.29	0.38	0.59	Lineal Foot
Welding Labor						
Joint	1.00	1.23	1.33	1.60	2.00	Each
Cap	0.73	0.89	1.00	1.33	2.40	Each
Elbow (45° or 90°)	1.60	2.00	2.29	3.20	4.80	Each
Flange—Slip-on	1.33	1.60	1.78	2.67	4.00	Each
Flange—Welding neck	0.80	1.00	1.14	1.60	2.40	Each
Reducer—Eccentric	1.45	1.78	2.00	2.67	4.80	Each
Reducer—Concentric	1.60	1.78	2.00	2.29	4.00	Each
Tee—Straight	2.67	3.20	4.00	5.33	8.00	Each
Tee—Reducing	3.20	4.00	4.57	5.33	8.00	Each

Table 2-5   Estimate Work Sheet

	Rate (Hours)	Qty	Extension (Hours)
Pipe Handling	0.59	21 feet	12.4
Elbow	4.8	2 each	9.6
Tee	8.0	2 each	16.0
Flange	2.4	2 each	4.8
Reducer	4.8	2 each	9.6
		Total	52.4

Construction Estimates Using Scaling

A quicker construction style estimate can be made using scaling. A repetitive task is estimated once and then multiplied by the number of times the task is performed. For the previous example, a standard estimate is applied to each weld. Assume that a weld on 6-inch Schedule 40 pipe takes about 2.7 hours to complete and a weld on a 3-inch Schedule 40 pipe takes 1.3 hours to complete. The planner develops the estimate after making a tally of the required welds, as shown here:

14 welds on 6-inch pipe at 2.7 hours per weld = 37.8 hours

2 welds on 3-inch pipe at 1.3 hours per weld = 2.6 hours

Subtotal Welding = **40.4 hours**

30 percent to prepare pipe = 0.3 × 40.4 hours = 12.1 hours

Total Labor = **52.5 hours**

There were 14 6-inch welds, and two 3-inch welds. An extra 30 percent of the welding time is added to account for the time required to cut, bevel, and lay out the pipe. The pipe required for the job must still be estimated using the take-out calculation.

Scaling can be applied to many repetitive tasks. It's a good idea for a planner to develop standard estimates for these tasks. Examples of where this approach can be applied include the following:

- Time per weld for pipe of different size, type, and schedule
- Cleaning time per tube in a heat exchanger
- Painting time per square foot of wall space

Usually, extra time must be added to the estimate for set-up and cleanup time.

Professional Planning and Scheduling Organizations

The following professional organizations can be a help to those performing construction (as well as maintenance) estimates:

* *American Association of Cost Engineers (AACE), International*—209 Prairie Ave., Suite 100, Morgantown, WV, 26507-1557, (304) 296-8444, www.aacei.org.

    The AACE is for cost engineers involved in cost estimating, scheduling control, project management, and related areas. This organization offers a certification program in Certified Cost Engineer and Certified Cost Consultant. A monthly magazine, *Cost Engineering,* resources, *Cost Engineers Notebook* (with updates), and *Recommended Practices and Standards* are offered with the annual membership fee.

* *American Society of Professional Estimators*—11141 Georgia Avenue, Suite 412, Wheaton, Maryland, 20902, (301) 929-8848, www.aspenational.com.

    The ASPE is dedicated to the advancement of estimators and the development of standard practices within the estimator profession. A certification is provided, called Certified Professional Estimator. The organization publishes a bimonthly magazine called *The Estimator.*

Another common estimating method, method time measurement (MTM), actually had its roots back at the turn of the twentieth century.

Methods Time Management (MTM)

The early industrial engineers considered the human body as a latent source of labor power. As such, the efficient and productive use of this labor power was deemed essential to increasing production and reducing costs. Building some semblance of science into the work process seemed to be logical and imperative if productivity improvement was to be achieved.

Lillian Gilbreth (1878–1972) and her husband Frank (1868–1924) are credited with first developing the field of scientific management. The Gilbreth's family of 12 and their work were depicted in the book *Cheaper by the Dozen* (New York: HarperCollins, Perennial, 1949), which was later adapted into a popular movie. Frank gained an interest in the connection between efficiency and standardization as an apprentice bricklayer. While observing two master bricklayers, he came to the conclusion that each man performed the same job differently. He became determined to find out which way was best.

Beginning in 1903, Frank and Lillian analyzed the motion of industrial workers, baseball players, physicians, homemakers, and others. Each movement of the body was broken into 16 basic elements they called *therblig's* (Gilbreth spelled somewhat backward). They believed that the one best way to perform any task could be found through observations called motion studies, which divided all activities into a series of therblig's. The time saved through more efficient use of the body was hoped to create what they called happiness moments for the worker.

Fredrick Winslow Taylor (1856–1915) has become known as the father of scientific management, although he actually created a system that combined Gilbreth's motion studies with precision time measurement of the basic elements of movement. With *time-and-motion studies,* the worker's happiness factor was ignored and the new goal was increased profits for factory owners. Taylor and his approach to scientific management came to the attention of the general public when his methods were used to resolve the Eastern Rate Case of 1910. Railroad companies had petitioned the Interstate Commerce Commission to increase the fee they were allowed to charge for transporting materials around the country. Louis Brandeis, a Boston lawyer who later became chief justice of the U.S. Supreme Court, used Taylor's scientific management methods to fight the rate hike.

With the help of a number of Taylor devotees (such as Frank Gilbreth, Henry L. Gantt, and Harrington Emerson), he was able to prove using time-and-motion studies that the railroads could be more efficient, with a potential savings of a million dollars a day. This savings would eliminate the need for a rate hike. When the news got out, the Taylor method was in demand throughout the world.

In the same year of this success, Taylor's method was hit with a major setback. The Watertown Arsenal in Massachusetts was a U.S. government-operated munitions manufacturer. In 1910, the molders at the plant went on strike against what they called stopwatching, performed by Taylor consultants. As strikes go, this was a minor event, but the fact that it was against a strategic government facility raised interest in Washington, DC. A congressional hearing into the strike and Taylorism was held in 1912. Taylor was forced to defend his methods against the notion that they dehumanized the work force and treated them like machines. Nothing of legal substance ever came of the hearing, but the running fight between organized labor and management had a new battleground.

Taylor was demoralized by the congressional hearings. He died three years later. However, the Taylor method lived on and scientific

Table 2-6 Time Measured Units (TMUs) for Reach Activities

Reach	
Distance	*TMUs*
0 to 5″	7
6 to 10″	13
11 to 20″	20
21 to 30″	31

management permeates the industrial landscape even today. In the first half of the 1900s, industrial engineers, in the mold of Taylor, began to review the assembly processes that utilized *piecework*. Piecework meant each worker was responsible for assembling only one piece in the manufacturing process—management was responsible for the whole process. Using what was called methods time measurement (MTM), the elemental activities of the worker (no longer called therbligs) were recorded and were timed in *time measurement units (TMUs)* of one-hundred-thousandth of an hour (0.00001 hour = 0.0006 minute = 0.036 second). Activities such as reach, move, grasp, disengage, and so on, are analyzed and timed. Table 2-6 and Table 2-7 are examples of these measurements.

Changes were made in the way assembly line workers were positioned, how they moved their bodies, and the way their hands and arms were moved, based on analysis of the elemental moves. The result was better eye–hand coordination and a reduction in manufacturing time. Recent medical studies have concluded that

Table 2-7 Time Measured Units (TMUs) for Move Activities

Move	
Type	*TMUs*
Walking	6 per foot
Sitting	30
Kneeling	80
Bending	40
Foot Motion	1 per inch

some of these repetitive and mechanical activities lead to muscle, nerve, bone, and tendon disorders such as carpal tunnel syndrome and residual physical stress. Industrial engineers have responded with ergonomics, an adjustment of the basic movements to benefit the health of the employee.

MTM and Maintenance Work

Since the introduction of scientific management into the production process, many attempts have been made to apply elemental movement techniques to maintenance work. This effort was problematic because most maintenance work is not repetitive like piecework. Also, similar jobs may require different tools or methods to perform them. Unbolting a man-way can require a wrench, or it may require a cutting torch (if the bolts are rusted in place).

The application of MTM and TMUs to maintenance work lead to the development of *engineered time standards*. Elemental activities were applied when analyzing basic tasks such as turning a bolt, hammering a nail, or cutting a pipe to generate the next elemental level of work. Table 2-8 shows an example of such measurements.

For example, if a 215-inch cut were to be made in a 1-inch thick piece of steel plate, the time required for the cut would be,

$$215 \text{ inches} \times 0.00200 \text{ hour per inch} = 0.43000 \text{ hours}$$

The task would be given a computer ID of CS13215 (CS13000 plus 215 inches) to help retrace the source if the estimate had to be audited.

This task is then made part of a much larger job, such as fabricating and installing a new man-way cover on a tank. The task may be incorporated into a job as shown in Figure 2-3.

Table 2-8 Engineered Time Standards for Flame Cutting a Steel Plate (Square Burn)

Plate Thickness	ID Plus Inches	Hours per Inch
$1/16$ inch to $1/4$ inch	CS10000	0.00110
$5/16$ inch to $1/2$ inch	CS11000	0.00143
$9/16$ inch to $3/4$ inch	CS12000	0.00173
$13/16$ inch to 1 inch	CS13000	0.00200
$1 1/8$ inches to $1 3/4$ inches	CS14000	0.00245
2 inches to $3 1/2$ inches	CS15000	0.00335

MTM Work Sheet

Job:	Fabricate and install new man way cover, 1" plate steel, 68" dia.				
Crew: 2 craft employees	**Planner:** Steven Smith				
Date: 5/10/91					
Element Description		**ID #**	**Unit Time (Hours)**	**Freq.**	**Total Hours**
Move O/H Crane 40"		CR42040	.01525	1	.01525
Clamp and hold 1" sheet steel		CR13000	.05250	1	.05250
Move O/H Crane and steel 30"		CR42030	.01144	1	.01144
Position 4 blocks to hold steel		MA62001	.04305	1	.04305
Position plate on blocks for cutting		PS33400	.10145	1	.10145
Mark 34" radius circle on plate		MP63340	.05345	1	.05345
Move cutting torch and tanks to work, 15''		DF98015	.02450	1	.02450
Flame cut circle in plate, 215" circum.		CS13215	.43000	1	.43000
Move scrap back to rack, 30'		CR42030	.01144	1	.01144
Move crane back to storage, 40'		CR42040	.01525	1	.01525
Fabricate rigging handles		WL60054	.11150	2	.22300
Weld on rigging handles		WL54385	.12835	2	.25670
Grind and clean edge		GS70215	1.43750	1	1.43750
Drill 1" bolt holes in plate		DH20111	.02550	54	1.37700
Move new manway cover to the tank		TM50018	.39000	1	.39000
Rig and position cover on tank		PS70150	.46730	1	.46730
Bolt cover in place		BF60100	.04000	54	2.16000
				Total	7.06983

Figure 2-3 MTM worksheet.

The total job is estimated to take approximately 7 hours. The job steps are broken down into identifiable elements that may mean a lot to the person developing the estimate, but mean little to the person performing the work. It is not the intention of this estimating method to give a welder a step-by-step approach to performing the job. Rather, MTM-based estimates are used for developing the hour estimate for the job.

The estimated job shown here can be grouped and added to a database of other jobs. Jobs estimated in this way are sometimes called *benchmark jobs* (not to be confused with a *benchmarking process* used to compare similar industrial facilities). This database can be queried by a planner/scheduler to determine the benchmark that best fits the job at hand.

Disadvantages of MTM-Based Estimates
MTM-based estimates do not take delay into account. Delay in MTM is considered to be the time is takes to do the job minus the

standard time set by the benchmark. This fact unwittingly draws companies into a mode of management that can have a long-term detrimental effect. In some companies, maintenance workers are made to fill out a delay card for each job. This quickly becomes a bone of contention between the management and the worker because blame for delays invariably is pushed down to the maintenance worker. A maintenance worker is less inclined to report the truth if the consequences could affect their performance evaluation or paycheck. A mechanic who notes that part of the delay on a job included 30 minutes waiting for the supervisor to get off the phone would surely feel awkward returning a delay card back to the supervisor. The mechanic would be equally reluctant to record the fact that he/she spent 5 or 10 minutes talking to a friend about fishing.

Some consulting firms provide a database of predefined estimates (benchmarks) to reduce the start-up time using the MTM method. These estimates are said to be independent of the bias normally associated with company-defined estimates. As a result, comparisons to standards (such as effectiveness and utilization factors) are claimed to be a true indicator of the quality of the work force or management. The jury is still out on this claim.

If a sufficient benchmark does not exist, a new benchmark must be built from scratch. It is unrealistic to believe that most planners would return to the elemental activity tables to build new estimates. What usually happens is that the planner chooses a benchmark that is close enough for practical purposes. Additionally, it is not uncommon for planners who are required to use MTM benchmarks to fudge the system by consistently using a combination of 10 to 25 different benchmarks out of a potential pool of 30,000 to 50,000 on file.

Additionally, an MTM-based process fails to identify needed materials, parts, tools, and equipment as other estimating methods do. A planner looking through a database of jobs may choose a job that is close to the job at hand, but the material and equipment required for the job must be inferred by the planner. A step-by-step review of the job is more likely to bring the required material and equipment to light.

The estimate developed through MTM is not usable as a step-by-step guide to the person performing the work. The job steps have little meaning to the maintenance employee. As a result, the maintenance employee is required to use a personal approach to the job, or a planner must build a separate description of the job steps.

MTM-based estimates have very little to add to the estimating requirements of a modern organization. The attempt to take what worked in a piecework environment and use it in the maintenance field has had a generally negative effect wherever it has been applied. Most companies that bought into this idea in the 1950s and 1960s are trying to find a way to wean themselves from the effort, but old ideas die hard. Companies made a big investment in the initial database of benchmark estimates and training of the current planners. Large industrial engineering staffs have been developed at some companies with the intention of monitoring the data and calculating the performance, utilization, and effectiveness factors. These companies need to bite the bullet and abandon this system. The goal of a planning effort should be to provide an estimate, a list of tools, parts, and equipment required, but not a time standard to measure worker productivity.

The Planning Thought Process

One estimating method that uses the abilities of a planner to the fullest is the *planning thought process*. The planning thought process has three main purposes:

- Establish an estimate of time required for the job.
- Define understandable steps to complete the job.
- Identify the materials, parts, tools, and equipment required for the job.

An approach that yields the best result is to visualize the job and how it will proceed. *Visualization* means forming a mental picture of a job and the persons working on it. This tends to bring more aspects of the job to light (such as the need for tools, parts, or equipment).

A plan developed through the planning thought process differs greatly from one developed using MTM and other so-called engineered estimates. Steps developed through the planning thought process can be used by a maintenance worker as a guide through the completion of the job, if this is required. Additionally, planning thought process estimates are developed in more manageable, and some may say realistic, time units. The shortest duration for a job step in the planning thought process is 0.1 hours, as opposed to 0.00001 hours in MTM-based estimates.

One method that helps a planner to visualize a job process is the diagram of question shown in Figure 2-4. Breaking the job into steps and then asking these questions about each step helps in the

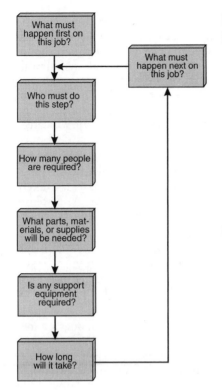

Figure 2-4 Planning thought process.

development of a better overall plan. This planning thought process is described as follows:

- *What must happen first on the job?*—While visiting the job site, a planner normally will ask what must happen first to begin the job. Does the operator have to be contacted? Will equipment have to be locked out? Will tools or parts have to be collected first? These are common first steps to a maintenance job.

- *Who must do this step?*—A planner will next identify the crafts or minimum skill level required to complete this first step. The skill required for one step might not be the same as those required for subsequent steps. Sometimes an overqualified person may be allocated to a few steps in a job to better expedite the total job, rather than switching people in the middle of the job.

- *How many people are required?*—While thinking about the job step, the planner may determine that more than one person is required to complete it. The first inclination should be to determine the minimum number of people required. If reduced downtime is the goal of the plan, the planner may opt to staff this step with two or more people. The maximum number of people required on the step will be limited by the space available and physical conditions, all of which the planner can easily assess at the job site.

- *What parts, materials, or supplies will be needed?*—Many planners do a very good job determining the labor requirements but miss the material requirements. This is less likely to happen if the job is broken down into steps. As a step is reviewed, the parts, materials, and supplies required will be obvious.

- *Is any support equipment required?*—Just as with material, special tools or equipment required for a job are more easily defined when the job is broken down into steps. Is a fork truck required? Is a torque wrench required? Is a pickup truck required?

- *How long will it take?*—Here is where the planner's experience is put to the test. Fortunately, when jobs are broken down in logical steps, elapsed time estimates for these steps will fall out naturally. The planner will usually multiply the number of people assigned to the job times the elapsed time estimate to determine the total labor hours.

- *What must happen next on this job?*—After all aspects of a specific step have been reviewed, the planner will identify subsequent steps. This thought process should be repeated to identify labor, material, and equipment requirements.

The planner should write down the steps as they are visualized. A simple planning worksheet such as the one shown in Figure 2-5 is used as an aid to put the thought process down on paper. This planning sheet is designed to help the planner through the steps of the thought process. It's filled out in the following manner:

1. The *W/O#* (work order number) and Equip. # (building or equipment identification) are used to tie the planning sheet to a specific job.

2. The *Seq.* (sequence) is used if multiple crafts exist. A planning sheet is filled out for each craft to be used, and the sequence

Planning Sheet

W/O #	Equip.	Seq.	Planner	Date
Job Title:				

Job Scope			Material, Tools & Equipment			
Job Steps	Crew	Est. Hours	Description	Qty	Stock #	Cost
Drawings/Forms:	Total Hours		Total Labor Cost $		Total Mat. Cost	$
					Total Job Estimate	$

Figure 2-5 Planning sheet.

(some work-order systems refer to this as a job step) ties all crafts to the original work order.

3. The *Job Scope* is where the thought process is noted. It is useful to write down each specific step and estimate the hours needed to do that step. Planners tend to estimate job hours in multiples of 4 hours. In other words, without rigorous estimating, job estimates tend to be 4, 8, 16, or 32 hours in length. This is sloppy estimating and removes any credibility from schedules in the effort to optimize resources.
4. The *Material* section allows for identifying any parts or materials needed. As the job scope is being developed and needed parts come to mind, it is best to note them immediately.
5. Any special *Tools and Equipment* are noted. This could entail special hand or power tools, alignment kits, or in-house support equipment (such as forklifts).
6. Any *Drawings/Forms* can be referenced. Many planners will use the backside of the planning sheet for a simple sketch and incorporate the planning sheet itself as part of the work order.

When the planning sheet is complete, it can be used as an input document for a computerized maintenance management system or simply filed for future reference.

A well thought-out planning sheet can be used to develop a plan for most job types. Following is an example of how this sheet can be used.

Planning Sheet Example

A recent PM of the West Dust Collector Blower indicated that the belts were cracked and worn. The maintenance supervisor wrote a corrective action work request to replace the belts. The planner began planning the job by visiting the job site. The estimate shown in Figure 2-6 was developed.

The estimate was developed in seven steps. The following explains the thought process of the planner:

1. First, the planner realized that it would take some time to receive instructions, get the materials required for the job, and go to the job site. The planner also realized that the motor starter for the blower would have to be locked out of service and safety-tagged. The mechanic doing the job would also have to discuss the equipment shutdown with the operator, which is part of the lock-out procedure. As a result, the mechanic was allowed 0.5 hours to perform this step.

2. The planner then noted that the belt guard will have to be removed. The planner has chosen to have the mechanic remove the belts in this step by loosening the motor mounts and jacking the motor closer to the blower. The mechanic may also choose to just cut the old belts off. Either way, the motor will have to be moved closer to the blower to get the new belts on, so an estimate of 0.25 hours is assigned to this task.

   The material, tools, and equipment column of the planning sheet helped the planner realize that some combination and socket wrenches will be required to complete this step. The planner measured the bolt heads and nuts at $9/16$ inch and $1/2$ inch. Other wrenches may be needed for other steps so the planner writes a reminder for the mechanic to bring a complete set of wrenches to the job site.

   Other planners may decide not to identify the wrench requirement. These planners assume the mechanic knows wrenches are needed before leaving the shop and will bring them. This is a reasonable assumption if mechanics have a tool pouch or small toolbox that they bring with them on any job. If this is not the case, it's a good idea to note all tool requirements to cut down on travel time.

3. Since the belts are damaged, the planner suspected the sheaves might be worn or damaged as well. A visual inspection of the

Planning Sheet

| W/O # 24585 | Equip. West Dust Collector Blower | | Seq. | Planner TMH | | Date 5/22/02 |

Job Title: Change V-belts. Cracked and worn.

Job Scope / Material, Tools & Equipment

Job Steps	Crew	Est. Hours	Description	Qty	Stock #	Cost
	1 ME					
1. Lock out and tag motor starter		.5				
2. Remove belt guard, loosen, and jack the motor. Remove belts.		.25				
3. Check both sheaves for sidewall wear, radial runout and wobble. Runout and wobble should not be more than .015" for the large sheave and .010" for the small. Replace worn sheave or hubs if required.		.5	Dial indicator, mag base (tool crib)			
			Sheave gauge (tool crib)		If needed →	
			If sheaves are worn:			
			4MV5V150R sheave	1 ea	43-515-100	$173.00
			1 7/8" bore R1 bushing	1 ea	43-923-750	24.00
			4MV5V109R sheave	1 ea	43-510-100	124.00
			2 3/8" bore R1 bushing	1 ea	43-918-750	24.00
4. Install new belts, align sheaves, tighten motor bolts.		1.0	5V120 belts	4 ea	06-510-112	53.00
5. Tension belts using force deflection method and recheck alignment.		.5	Belt laser alignment tool			
			Belt tension gauge (tool crib)			
6. Re-install guard.		.25				
7. Remove locks, release to operations and clean up.		.5				

Drawings/Forms:	Total Hours 3.5	Total Labor Cost $130	Total Mat. Cost $53.00
			Total Job Estimate $183.00

Figure 2-6 Planning sheet example.

sheaves is augmented with the use of a sheave gauge that can aid the mechanic in identifying worn sidewalls in the sheave. The sheaves may also be out of round (run-out) or cocked on the shaft (wobble), both of which can be measured with a dial indicator mounted on a magnetic base. Both of these tools are noted in the tools column as well as the fact that they are stored in the tool crib. The planner noted that there is a plant standard for the indicator measurements on sheaves and left some space to put this information on the sheet. These blank spaces will be filled in back at the office. An estimate of 0.5 hour is given for this step.

In the unlikely case that the sheaves are worn or that severe wobble or eccentricity is present, the planner makes a note to identify the sheave size and hub size. These records are back in the office along with information about the belts required. The estimate of 0.75 hours is not sufficient to perform the sheave replacement. However, the likelihood of this happening is small, so the planner decides to include only the belts.

4. The next step is the installation of the new belts. The sheaves must be aligned with a new laser tool, which should be removed from the tool crib and brought along to the job. The belts should be pretensioned before the tension gauge is used, so the planner noted this fact. Installing the belts, aligning the sheaves, and pretensioning the drive system will take about 1 hour.

5. The belt tension will be checked as the jacking bolts are turned and after the motor bolts are retightened. This process may have to be repeated three or four times to do it right A spring-type belt-tension gauge is used to check the tension, which is also available in the tool crib. The planner leaves it up to the mechanic to read the instructions for this gauge to properly tighten the belts, rather than spelling out the proper tension. This step is estimated at 0.5 hours.

6. Another 0.25 hours is assigned to reinstalling the belt guard.

7. Removing the locks, returning the equipment to operations, and returning to the shop for a new assignment will take about 0.5 hours. After reviewing the steps in the job, the planner returned to the office to finish the estimate.

Back in the office the planner reviewed the plant records that have the sheave diameters, as well as the hub size and type for both sheaves. The belt size is also noted along with the fact that four are required. The stores part numbers are looked up and recorded for both of these items. The planner checked the stores database to see if the parts were available. If they were available, the job could have been scheduled in the next open slot. If the parts were not available, a request would have been made to the storeroom to order these items.

The planner totaled the estimate (3.5 hours) and calculated the cost of the job based on a labor rate of $37 per hour. The total job will cost about $183.

The planner can pull the belts out of the storeroom ahead of time if the storeroom is not on the way to the job. The mechanic should withdraw the tools from the tool crib.

A planner may choose to attach a copy of the plan to the work order. Some planners may decide just to list the parts and tools required along with the standards for sheave run-out and wobble on the work order. An example of such a work order is shown in Figure 2-7.

Maintenance Work Order					#45672

Date Initiated	Originator	Downtime?	Date Available	Time Available	Priority
05/17/02	SMITH	YES - EQUIP. ONLY	05/30/02	08:00	PLANNED

Equip. #	Equipment Description	Date Required
BL-203W	WEST DUST COLLECOR BLOWER	05/30/02

Classification: **CORRECTIVE**

Work Requested: CHANGE BELTS.   CRACKED AND WORN.

INSPECT SHEAVES FOR DAMAGE.   USE SHEAVE GAUGE TO CHECK SIDE WALLS.

CHECK RUNOUT AND WOBBLE.

LARGE SHEAVE LIMITS .015" FOR RUNOUT AND WOBBLE

SMALL   "   "   .010"   "   "   "   "

Material & Parts Description	Stock #
4 EACH 5V1120 V-BELTS	06-510-112

Tools:

- SHEAVE GAUGE   - BELT TENSION GAUGE

- DIAL INDICATOR W/MAG BASE   - BELT LASER ALIGNMENT TOOL

Safety Checks:

CONTACT OPERATIONS BEFORE STARTING WORK

LOCK AND TAG BEFORE BEGINNING WORK

Comments:

Figure 2-7 Maintenance work order.

This work order may be the only thing the mechanic needs to perform the job. If the sheaves must be replaced, the mechanic can contact the planner for the sheave data.

Note in the previous example that the planner went to great lengths to limit additional trips back and forth to the storeroom or tool crib. This is the primary area where the planning effort pays for itself.

Economics of Planning

In many maintenance planning efforts, it is wrongly assumed the jobs that occur frequently do not need any planning effort at all. The conclusion is made that these jobs are performed so often that the maintenance worker should know what must be done and what is needed to do it. However, it is precisely these jobs that are frequently interrupted by trips from the job site to get a tool, or trips to and from a storeroom as additional materials and parts are required.

The effectiveness of a planning effort is based on economics. When the function of planning is defined within an organization, it is assumed that the planner's effort in reviewing work ahead of time will be more than recovered in efficiencies when the job is eventually performed. Equipment downtime and lost labor time can be reduced by ensuring that materials, tools, and equipment are made available before the job starts, not during the job.

With this in mind, what kind of work should a planner spend time on? Consider a plan for a job that does not involve equipment downtime and that occurs about once a year. The probabilities of a return on this planning effort are low unless the job is also long and complex. Savings are even questionable for a job that will occur on an average of twice a year. Even so, more savings can be expected, because the job is being performed more often using the same plan.

However, the economies begin to change when a job is considered that occurs about once a month. The savings achieved through planning can be low for each time the job is performed, but the overall savings accumulate through the frequency of occurrence. Simply, one hour saved per month on a job that took only one hour to plan can return a twelve-fold return in a year.

Jobs that occur more frequently should be considered first in a planning effort for another reason. These jobs fall into the category that is often referred to as *generic maintenance*. Generic maintenance is work that doesn't change much from industry to industry, and the job steps do not change with the size of the equipment. Consider the previous job of changing V-belts on a small blower. The belts were 5V1120 in size, indicating a belt circumference of 112 inches. Even with 4 groove sheaves, the motor size in such an installation would probably not exceed 20 hp, and the shaft centers would not be separated by much more than 3 feet. One mechanic can perform the job easily.

Now consider another blower that is driven by a 150-hp, 8 groove sheaves employing 8V4500 belts. In this installation, the belt guard must be lifted by a cherry picker and the whole job will require at least two mechanics. However, the job steps involved in changing these belts are the same as for the smaller motor. Of course, the actual resources needed at each step will change, but these resources are much easier to identify and quantify when a list of job steps already exists. In other words, one plan can be used to build another more quickly.

Job plans generated for generic maintenance work can also be used to plan out work that may involve some unusual aspects. A job plan developed for changing out a 50-hp motor located at ground level with easy access could also be used as a basis for a plan to change out a motor located down in a pit. The job steps remain relatively unchanged. The resources required to accomplish many of the steps will probably change. For example, the initial step of gaining physical access to the motor will involve more than a lock-out procedure. Now a confined-space entry procedure will also be needed and a safety watch individual may have to be added to the crew.

Developing Generic Estimates
It is advisable to develop a list of standard job estimates. These standards can be adjusted after analyzing experience on actual jobs. Following are some categories of work for which standard estimates are developed:

- *Equipment change-outs*—Motors, pumps, fans, instruments.
- *Common field repairs*—Bearings, belts, chain, seal replacements.
- *Shop overhauls*—Pumps, gearboxes, instruments.
- *Piping installations*—Time/weld for welded pipe, time/joint for threaded or composite pipe.

Though it is important to go to the job site when developing a plan, the time spent at the site can be limited if the job to be performed is common to the facility. A preplanned job, using the planning thought process, can be easily modified to cover a similar (but not identical) job. A very good example of this is the replacement of a horizontally mounted motor. In the document shown in Figure 2-8, the planner has developed a generic job plan for a motor replacement.

Replace a horizontally mounted motor (1800 RPM and 3600 RPM) *						
	10 HP	20 HP	40 HP	50 HP	100 HP	200 HP
	1 Emp.	1 Emp.	1 Emp.	2 Emp.	2 Emp.	2 Emp.
Lock out and tag the motor starter	20 min.	20 min.	20 min.	40 min.	40 min.	40 min.
Electrically Disconnect	10 min.	10 min.	10 min.	40 min.	60 min.	90 min.
Uncouple, unbolt, and remove motor	10 min.	15 min.	20 min.	50 min.	90 min.	180 min.
Install new motor	25 min.	30 min.	45 min.	90 min.	120 min.	240 min.
Align to within .002"	90 min.	90 min.	120 min.	240 min.	300 min	420 min.
Electrically connect, check for rotation	25 min.	25 min.	45 min.	70 min.	90 min.	120 min.
Remove lock and return to operations	20 min.	20 min.	20 min.	40 min.	40 min.	40 min.
Total	200 min.	210 min.	280 min.	570 min.	740 min.	1130 min.
Total Labor Hours	3.3 hrs.	3.5 hrs.	4.7 hrs.	9.5 hrs.	12.3 hrs.	18.8 hrs.
Special Equipment	Fork Lift	Fork Lift	Fork Lift	Fork Lift	Crane	

* For lower RPM motors, use estimate for next higher HP

Figure 2-8   Generic estimates.

After developing a plan for a 10-hp motor, the planner went ahead and developed similar plans for higher-horsepower motors. Only the steps that would take longer were modified in the estimate. For example, it would still take the same 20 minutes for a maintenance worker to go to the job site and lock out the motor starter on a 20-hp or 40-hp motor. Installing these larger motors may take a little longer because of physical size, so the planner increased the estimate accordingly.

The planner also noted that two people would be needed on the job if the motor was 50-hp or above, again because of the physical size. The time required for lock-out and removing locks had to be doubled because both employees would have to place their locks on the starter (per plant rule).

The fact that a crane would also be needed for larger motors is also noted. Motors with more than 50 hp would have to be moved to and from the job site by crane rather than by a forklift with slings. Also, the planner realized that the original estimate was for an 1800-rpm motor and that a 3600-rpm motor would have a smaller frame size (true for most TEFC motors).

The visit to the job site can be used to ascertain additional information unique to the job at hand. Vibration measurements can be taken, or special conditions can be noted before developing the final plan.

Step Estimate Accuracy

The plan shown in Figure 2-9 was developed for an instrument job. This job scope was developed using only a few steps. The detail work is implied or the planner is depending on the expertise of the instrument mechanic. A calibration procedure (which is already developed) is referenced under Drawings/Forms rather than spelling out the details of the procedure. A dollar estimate of $70 was recorded as well. This estimate was based on a labor rate of $35 per hour.

No special tools are identified but the planner noted that only normal instrument mechanic's tools would be required. An instrument calibrator and special mounting rig were also identified so that the employee knows what to pull together. It's also a good idea to identify this equipment because it may be in limited supply (other instrument mechanics may be using them).

The shortest interval used for the instrument job described earlier was 15 minutes (0.25 hours). Figure 2-10 shows a more detailed plan of the same job, using shorter job steps. This estimate is four pages long; only the first page is shown. Each individual tool was identified. Some planners may feel that every tool should be identified because the employee may make return trips to the shop for tools. This tool detail is usually unnecessary if instrument mechanics normally carry tool pouches with them to the job (as they should).

The planner used 0.1 hours (6 minutes) as a minimum time interval for a step. Many of the individual steps identified will take less than 0.1 hours to perform, so the planner assigned 0.1 hours to a group of two or three steps. For example:

- *Move pointers and check for ease of operation*—No estimate
- *Connect instrument to calibration and alignment fixture*— 0.1 hours

Implied is that both moving the pointers and connecting the instrument to the calibration unit will take about 0.1 hours. The

Planning Sheet

W/O # 43762	Equip. IL420 - Hot Temp., Zones 4, 5, & 6	Seq.	Planner MAB	Date 7/27/91

Job Title: Calibrate Foxboro 110 Indicator

Job Scope

Job Steps	Crew	Est. Hours
Contact control room operators and have them switch to manual operation		.25
Remove instrument and bring it to the shop		.25
Perform calibration procedure INST0005		1.0
Repair, replace parts as necessary		.25
Reinstall instrument and inform operations		.25

Material, Tools & Equipment

Description	Qty	Stock #	Cost
Normal Inst. Tools			
Wallace & Tiernan air calibrator			
Foxboro C0138SN cal/align unit			

Total Hours 2.0
Total Labor Cost $70.00
Total Mat. Cost $
Total Job Estimate $70.00

Drawings/Forms: INST0005 Cal Sheet

Figure 2-9 Sample estimate.

Planning Sheet

W/O # 43762	Equip. IL420 - Hot Temp, Zones 4, 5, & 6		Seq.	Planner MAB			Date 7/27/91

Job Title: Calibrate Foxboro 110 Indicator

Job Scope

Job Steps	Crew	Est. Hours	Material, Tools & Equipment			
			Description	Qty	Stock #	Cost
Contact control room operators and have them		.25				
switch to manual operation						
Disconnect field connections		.1	#1 Screwdriver			
Remove indicator from shelf mount						
Move instrument to shop bench		.25				
Open side door of instrument			#1 Phillips			
Inspect receiver elements for damage or wear						
Repair, replace as necessary		.25				
Move pointers and check for ease of operation						
Connect instrument to calibration and		.1	Foxboro C0138SN cal/align unit			
alignment fixture						
Apply 3, 9, and 15 psi to input B		.1				
Record green pointer reading on calibration sheet						
Apply 3, 9, and 15 psi to input C		.1				
Record red pointer reading on calibration sheet						
Apply 3, 9, and 15 psi to input D		.1				
Record blue pointer reading on calibration sheet						
Apply 9 psi to receiver unit, square up pointer		.1	Wallace & Tiernan air calibrator			
and use zero screw to moving plate and			Instrument Screwdriver			

Drawings/Forms: INST0005 Cal Sheet	Total Hours Page 1	1.35	Total Labor Cost $		Total Mat. Cost	$
					Total Job Estimate	$

Figure 2-10 Sample estimate—more detail.

planner chose to use .1 hours for the combination of steps rather than breaking the estimate down even lower to 0.05 hours (3 minutes).

The total estimate is 1.35 hours (actually, the estimate for page one). This number implies an accuracy that does not really exist. Job step estimates of 0.1 hours have an accuracy of about ±50-percent, whereas a 0.25-hour estimate will have an accuracy about ±20 percent. A 1-hour job-step estimate has an accuracy of ±10 percent. A higher number of 0.1 hour estimates can contribute to a lower overall accuracy.

The graph in Figure 2-11 shows the result of one study on the accuracy of job steps using the planning thought process. The graph shows that the accuracy of a step gets worse when the step is small (such as 0.1 hours, which has an accuracy of about ±50 percent). The accuracy improves to ±10 percent for 1-hour step estimates, but gets worse again for higher step estimates. An 8-hour step estimate has an accuracy of more than ±35 percent. The accuracy drops off for higher step estimates because the likelihood of shotgunning (or guessing increases when the step estimates are large).

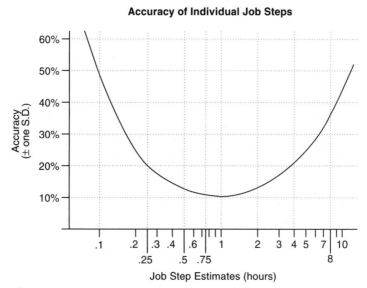

Figure 2-11 Job step accuracy.

Table 2-9 Job Plan Accuracy

Short Plan		Detailed Plan (First Page Only)	
0.25 hrs × 20% = 0.05		0.25 hrs × 20% = 0.05	
0.25 hrs × 20% = 0.05		0.1 hrs × 50% = 0.05	
1.00 hrs × 10% = 0.1		0.25 hrs × 20% = 0.05	
0.25 hrs × 20% = 0.05		0.25 hrs × 20% = 0.05	
0.25 hrs × 20% = 0.05		0.1 hrs × 50% = 0.05	
		0.1 hrs × 50% = 0.05	
		0.1 hrs × 50% = 0.05	
		0.1 hrs × 50% = 0.05	
		0.1 hrs × 50% = 0.05	
2.0 hrs	**0.30 hrs**	**1.35 hrs**	**0.45 hrs**

Using the information shown in Table 2-9, we can tabulate the accuracy of the short and long version of the job plans.

$$\text{Accuracy} = \frac{3 \text{ hours}}{2.0 \text{ hours}} = \pm 15 \text{ percent}$$

$$\text{Accuracy} = \frac{45 \text{ hours}}{1.35 \text{ hours}} = \pm 33 \text{ percent}$$

A series of inaccurate steps reduces the total accuracy from ±15 percent to ±33 percent. The accuracy implied by the extra numbers to the right of the decimal position for the 1.35-hour estimate is obviously not really there.

A planner is advised to keep job steps estimates between 0.25 and 1 hour when using the planning thought process to estimate a job. The accuracy of the resulting plan will usually end up being below ±15 percent.

Delays caused by paid breaks, early quits, late starts, and traveling to and from breaks are not included in some planners' estimates. Ways to handle these delays are discussed later in this chapter in the section called *Handling Delays in an Estimate*.

Estimating Using Past Performance
Estimates based on historical data tend to be more consistent with actual activity. Once the job has been performed enough times, the

time required to develop historical estimates is relatively short compared to other methods. On the other hand, historical estimates do not necessarily reflect improvements in methods. Mistakes of the past become part of the norm if the poor plans of the past are used.

The job history shown in Figure 2-12 is printed from a computer database. This job was performed 19 times in the last 8 years. Each time it was performed the actual labor hours were entered into the database. The results suggest the job will take about 24 hours whenever it is performed. The computer program calculated the standard deviation (SD) to be about 3 hours. A planner can use this value as an estimate for the job because the accuracy is relatively high (±13 percent).

Job: Change Leaking Gas Seal on Scrubber Fan

W/O #	Date	Act. Hours
34689	3/17/84	27
36212	9/13/84	31
37735	2/6/85	20
39258	7/9/85	24
40782	1/18/86	21
42304	6/15/86	26
43827	11/17/86	24
45350	6/4/87	25
46873	11/5/87	25
48396	4/21/88	21
49919	10/15/88	19
51442	3/2/89	26
52965	7/5/89	20
54489	12/21/89	22
56011	5/20/90	22
57534	10/14/90	24
59057	5/7/91	27
60581	9/7/91	23
62103	1/30/92	28
	Average	24 hours
	SD	3 hours
	Accuracy	±13%

Figure 2-12   Job history printed from computer database.

Estimates based on past performance are best used for the same job performed on the same equipment, rather than the same job on different (but similar) equipment. Special conditions may exist on similar jobs for different equipment in a facility. For example, a pump replacement in one part of the plant may take about 8 hours, but a similar pump in another part of the plant may take longer to replace because of its physical location.

Problems with Estimates Derived from History

Estimates based on history or past performance can become the planner's most valuable resource in daily planning. The thoroughness and quality of such a resource is often hampered by inaccurate information that ends up in the historical record or important information that is excluded from the records. Following are some of the more common problems that must be corrected in work order history:

- *Completed hours charged to historical records may be inaccurate*—Work orders are often closed out with the completed hours charged to jobs that have little or no relation to the actual hours that were required on the actual job. If the job was an emergency, the work might have been begun before an actual work order existed to cover the emergency itself. The early hours expended to deal with the emergency were charged to another job and a correction to those jobs was never made. In another instance, a large job may have been used to cover any number of small jobs that were worked without orders to cover them. At the completion of the large job, the hours charged against that job are much more than was actually needed to do the work.

- *History records may not document all parts used on the job*— If an emergency comes up, parts are often needed before a work order has been created to cover the job. In such a need, the parts are often charged against a cost center or general plant charge account or even another work order. Even after a work order is created, the proper accounting corrections are not made to see that the work order is charged with all parts used on the job.

- *History records do not capture other resources*—Historical records will never capture resource requirements such as support equipment, special tools, free issue items, special rigging supplies, or any other resource that are not necessarily charged against the job but are still necessary in getting the work done, unless some special reporting procedure is in place to see that such resources are itemized as they are identified.

- *Historical records may not capture outside services*—In many plants and facilities, outside resources are often brought in under a blanket purchase order. The costs of these outside

resources will not reflect back to a given job or piece of equipment for which they were brought into the plant.

- *Closure notes on historical records do not clarify actual
  work performed*—When work orders are closed out after
  completion, typical closure comments might consist of done,
  completed, or worked as requested. Such comments provide
  minimal insight into the work performed or any special or
  unusual problems that might have been encountered during
  work execution.

Improving the Historical Record

With the previous shortcomings of historical records having been
noted, the following actions can be taken to vastly improve the
quality of historical records and improve the daily planning effort
that is made from past performance:

- *Write work orders for all maintenance activities*—Although
  writing work requests for all maintenance activity would ap-
  pear to be a clerical burden, there is no better means of ensuring that an accurate record of each job (including the hours of
  actual labor expended on them) is captured in history. Many
  plants and facilities, in an effort to develop paperless systems,
  put the onus of the generation of all work requests on the re-
  questor. This is beneficial in that when emergencies arise, the
  individuals that desire emergency assistance must generate a
  request for such assistance before any action can be taken on
  the part of maintenance.

- *Charge parts and materials to work orders only*—One way
  to capture all parts and materials used on a particular job or
  work order is to allow those resources to be chargeable only
  to a work order in the first place. This will not ensure that
  parts and materials are charged only to the correct work order,
  however. Many plants do not desire the rigidity of the previous
  procedure. If a cost center or general charge account is to be
  made available, it is a good practice to access this temporary
  account on a regular basis (daily if necessary) to remove all
  charges made into it by accounting for those charges against
  the actual work that required them.

- *Charge outside services to a work order*—Work orders can be
  used to track outside resources. Most CMMS programs will
  ensure that such costs are charged against the proper equipment, as long as that equipment has been specified on the work
  order itself. Work orders are ideal for tying together all charges

outside of plant resources and expended against maintainable assets.

- *Provide a means to capture undocumented resources*—Some planners utilize a closeout form to capture resources such as tools, support equipment, and any other resources that would ordinarily not be captured at the time that a work order is closed out. A typical form is shown in Figure 2-13.

Work Order Closeout

W/O # _____ Date _____/_____/_____

Support Equipment:
_____ Forklift _____ Carrydeck _____ Crane

Rigging Equipment:
___ Shackles   Size _____ Qty. ___
___ Chokers   Size _____ Qty. ___
___ Web Slings   Size _____ Qty. ___
___ Cribbing   Size _____ Qty. ___

Safety Equipment:
___ Personnel Locks   Qty. ___
___ Valve Locks   Qty. ___
___ Blinds   Size _____ Qty. ___
                    Size _____ Qty. ___
                    Size _____ Qty. ___

Free Issue Items:
Item _____ _____ Number Reqd
Item _____ _____ Number Reqd
Item _____ _____ Number Reqd
Item _____ _____ Number Reqd

Figure 2-13 Work order closeout form.

- *Encourage meaningful closing comments*—Enlisting meaningful comments on the work performed requires creative thinking on the part of planners. If the work order has requested that some form of documental procedure be used, then the crafts persons might be required to document final readings. As an example, the final alignment readings whenever motors or other drivers are replaced on direct driven machinery would be meaningful information in the history record.

- *Use work orders to update equipment database information*—The building of equipment database information in a CMMS is often a major effort for many plants and facilities. Upkeep of such information, however, is often overlooked and such databases are, in time, full of inaccuracies. Planners can begin to update the database by printing out current listings of a given piece of equipment when work orders are written against them and then having the people performing the work check the printed listing against the actual field equipment for corrections and updating.

Some CMMSs have a question on the closeout screen for a work order to facilitate historical estimates. Questions such as, "Should this estimate be added to the estimate history file?" are answered by the planner as the job is closed out. If the job was performed as planned, the planner would answer "Yes." If the work and charges were unusual or the maintenance supervisor used a larger crew than was necessary on the job, the answer would be "No."

Other systems require the planner to evaluate the data after reviewing each completed work order. Jobs that were canceled or jobs that had zero hours charged should not be included in the history to be evaluated. It's also a good practice to throw out the highest and the lowest actual labor hours to improve the distribution and accuracy of the estimate.

Short jobs (1 to 4 hours long) tend to provide less accurate historical estimates. This is mainly because supervisors tend to round up to the nearest whole number. Jobs taking less than an hour to complete are entered as taking 1 hour. Inaccuracies also occur on jobs lasting slightly less than 8 hours. Single craft jobs taking 6.5 to 8 hours are usually entered as 8 hours because no chargeable work was performed by the employee during the remaining time.

It often is a good idea to plot the data from past job performance to determine if the effects of *learning* are affecting the calculated average. The chart in Figure 2-14 shows data for a job that illustrates this point. Early job performance was poor because a good approach to the job had not yet been established. This job took 28 labor hours the first time it was performed and 32 labor hours the second time. Performance improved over the next three times it was performed. An average estimate of about 24 hours would initially be assumed by a planner if the data from the first five times the job was performed were used. The accuracy of this estimate is very low (about ±25 percent) because the distribution of the first five tries is very wide.

Decreasing job duration reflects the result of learning an improved method to perform the job. The average time for the remaining tries is about 14 hours, and the accuracy of the estimate is also improved. A conclusion can be drawn that the first few performance data points should be discarded when determining an estimate from historical data.

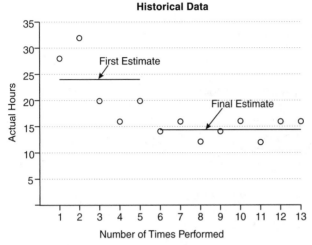

Figure 2-14 Estimating using history.

The Estimating Learning Curve

The key to properly estimating maintenance work is developing an effective schedule and ensuring the backlog is valid. The goal of estimating is to predict all the requirements of a job. This is possible to a point, but first attempts at planning often result in the actual job taking longer or shorter than the original estimate.

The best improvement in estimates comes from comparing job plans to the actual work-in-progress. As the work is being performed, a planner will compare what is actually being performed to what has happening in the field or on the floor. Specific questions such as the following are asked about each job step that deviates from the plan:

- What materials, tools, or support equipment were not included in the resource list?
- What job steps took longer than estimated?
- What job steps were completed ahead of time?
- Was the overall work plan enhanced or inhibited by the deviations?

An adjunct to inspecting the work-in-progress is to review jobs after the work is completed. This is especially beneficial after shutdowns, turnarounds, or outages when the pressure of the schedule precludes the luxury of inspection while the work is going on. Supervisors and crafts people should be involved. Go over the important jobs that will be performed again in the future and ask questions to fine-tune the job plan.

Such a review of efforts involves an important principle in estimating called the *learning process*. Whenever a particular job is repeated, one's ability to perform that job should improve. This assumes that an effort is being made on the part of the individuals involved to improve on the last attempt. A plot made of the time required for successive attempts at the same job is called the *learning curve*. The learning process is measured by developing a learning curve, as shown in Figure 2-15.

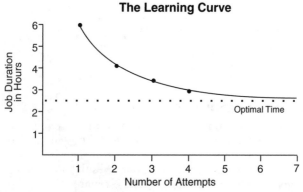

Figure 2-15 The learning curve.

The first organized use of a learning curve was developed in 1925 at Wright Patterson Air Force Base. A government accountant noticed that labor hours required to build the first aircraft were reduced to about 80 percent for the next order. Subsequent orders resulted in a further drop to 80 percent of the previous order. Learned methods and systems were cited as the source of the improvement.

The following example examines the construction of a Ferris wheel by a traveling carnival crew.

The first time the Ferris wheel is built, a number of components are assembled out of order. Sub-assemblies have to be dismantled and reassembled correctly. The supervisor of the assembly crew makes note of the mistakes and writes out steps for assembly to be used next time.

The next time the Ferris wheel is built, parts are discovered missing and must be rushed in or purchased locally. The supervisor develops a parts checklist for the dismantling and storage of the Ferris wheel.

In subsequent constructions, the carnival workers determine new methods and tools to speed the job and cut down the total construction time.

New technology may become available and the construction of the Ferris wheel may speed to a point where further improvement may not be possible or may be too costly. The time for construction will eventually approach 2.5 hours. This can be considered the optimum job time.

As we have learned in this chapter thus far, several factors can affect the accuracy of a job estimate. Organization of documentation can be key, assignment of the proper crew to a job can be crucial, and the preparation of a parts and materials requirements list can be essential.

Work Packages

Many planners collect all documents required to perform a job into a *work package*. A work package is a folder or group of folders that include all the necessary paperwork and reference material required to complete a job. An example of one such package is shown in Figure 2-16.

Work Package Contents   **Figure 2-16** Work package.

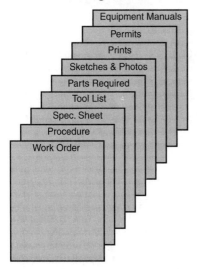

A copy of the work order should be included along with a written *procedure* of how the job should be accomplished. A *specification (spec) sheet* should be added for jobs that must meet plant standards. Spec sheets, which indicate the goal for the repair process, should also include some space for a mechanic to report as-found and as-left readings. An example of this would be an alignment spec sheet to show dial indicator readings before and after. *Tool* and *parts withdrawal slips* should be filled out ahead of time and included in the package. The planner should include *sketches* or *photos* of the work to be performed, and *prints* of the equipment or job site should also be included. All *permits* required to perform the work, such as line-entry permits and lock-out tags, should also be included. *Equipment manuals* can also be enclosed to help the maintenance worker through some tough procedures.

A work package may not be a requirement for every job, but it is very helpful to maintenance workers and first-line supervisors when working on shutdown, turnaround, and outage jobs.

Factors Affecting the Accuracy of an Estimate

Many factors affect the accuracy of the estimate. These include the following:

- Planning conditions
- Direct planner problems
- Insufficient time

Planning Conditions

The conditions that exist for planning are the first items to consider. Some of the most common of those conditions are the following:

- Degree of supervision
- Labor relations
- Facility condition
- Economy
- Weather

Degree of Supervision

Estimates will approach actual hours and most schedules will go off as planned if first-line supervision at a facility is well-seasoned. If this is not the case, the planner's estimates may not approximate actual times. The jobs in their charge may exceed the estimate.

Labor Relations

A facility that has strict craft lines poses a special problem to the planning and scheduling effort. First, the craft distinctions may require a separate plan for each craft. This is usually true at larger facilities. Second, if the different crafts are not under the same supervision, the planner must develop elaborate schedules to coordinate jobs. This will invariably lead to waiting time and delays between crafts. Some facilities divide the planning function up into separate craft lines as well. Close coordination between craft planners is important if all portions of the job are going to fit together properly. Here again, delays will be common when the craft requirement changes during the job's progress.

Some companies with craft distinctions have the ability to allow two different crafts to work together on the job. One individual can act as the lead and the other as the helper, depending on the craft requirements. This improves the estimating process, but perpetuates the buddy system, even for small jobs.

Other companies that have labored under craft restrictions are attempting to move toward a multiskilled effort. One craft employee may now be asked to perform a job that requires two or more craft skills. These employees are trained in the extra skills through formal training and on-the-job activities. The savings to a company in reduced delay time and production downtime can be as much as 25 percent. A deal is devised with craft employees at these plants to share in this savings through a pay increase.

NOTE

Multiskill training is discussed in more detail in Chapter 9.

Facility Condition

Old houses require more maintenance than new ones. This fact is also true for industrial facilities. Work on otherwise minor problems (such as the lights going out) often requires a major repair (such as pulling new wire or replacing a fuse box with a breaker panel). Even new facilities that were poorly constructed to begin with fall into this category. Maintenance in these facilities usually ends up fixing the engineering mistakes.

Economy

Changes in the economy mostly affect construction crews and not maintenance work. A good economy actually results in poor productivity from a construction crew. High-quality craft employees will be unavailable. The good ones already working for you may go

Table 2-10 Temperature Allowances

Temperature Allowances	Range	Adjustment
Hot	100 to 120°F	Add 20%
	120 to 140°F	Add 30%
	Above 140°F	Add 40%
Cold	0 to 20°F	Add 20%
	–20 to 0°F	Add 30%
	Below –20°F	Add 40%

elsewhere for more pay. If the economy is poor, good construction workers are plentiful.

Weather

Bad weather can be a detrimental factor on the overall performance of a work force. Extreme cold or hot conditions usually slow down work progress. Allowances can be made for certain *consistent conditions*. Table 2-10 shows some example of *allowances* for *high and low temperatures*.

Rain and *snow* are essentially unpredictable as far as planning and estimating work is concerned. No adjustment can be made for rain or snow in an estimate, but adjustments can be made on the day before the work is to take place. A planner may decide not to fill the schedule as much if snow is predicted for the next day.

Other Conditions

Special conditions may exist from job to job that should be taken into account. The following are typical adjustments made by planners.

- Work at heights on a ladder (over 12 feet)—Add 10 percent
- Work around moving equipment—Add 20 percent
- Work around energized electrical equipment—Add 20 percent
- Work in limited space—Add 20 percent

These values should be used only on the working steps of the job. Steps that include preparing for the job or locking out equipment should not be scaled up by these factors.

Direct Planner Problems

Direct planner problems can also be the source of problems in estimating. Sometimes the planner is not qualified to plan the job. Often planners are asked to plan a job in a field in which they have no expertise. Planners should be able to employ the knowledge of supervisors, skilled crafts people, or engineering to scope out jobs and identify needed resources. On large piping jobs, for example, many planners without a strong background in pipefitting will write a companion work order instructing a pipefitter to inspect the job, prepare a bill of materials, provide a sketch, and determine an estimate.

Insufficient Time

Insufficient time may be available to perform a good plan. If most of the maintenance work that is performed at a facility must be completed within the next few days, a reasonable plan may never be developed. Parts and materials will not be identified and the job time estimate will invariably be wrong. Scheduling will be very difficult under this situation and the only job that is started on time will be the first one of the day. Good plans require an average of *one week prior notice* to ensure the job will be completed near the estimated time and will not be delayed by lack of parts.

Assigning the Crew Size to a Job

The planner usually will make estimates based on the quantity of work required and the logistics involved when two or more employees are assigned to the same job. The goal of the planner is to always make sure that each employee assigned to the job is *gainfully employed* throughout the job. This is not always possible and some lag time will be experienced by one individual or another at any one time. This is particularly true during two-person jobs.

The responsibility for assigning specific individuals to the job often falls on the supervisor. The supervisor will make the choice based on who is best qualified to perform the work. Some jobs may present an opportunity for on-the-job training. The supervisor can assign the person who needs the training to the job when time allows.

Parts and Material Requirements

Material shortages and the failure of the planner to identify materials to be used on the job are major contributors to delays of work in progress. Some jobs must be stopped and rescheduled because a planner forgot an important part or just left it up to the supervisor to handle.

Other planners find their jobs have been reduced to that of a buyer or expediter. Many parts that are not held in the storeroom must be purchased and received by the planner. This is more often the case with people who plan instrument or electrical work in which nonstandard items are often required, but should seldom be the case for other items.

It is definitely important that the planner attain the parts. However, this effort should not take most of the planner's time. Often, working with the storeroom is the best answer. Planners should investigate the possibility of making the following improvements.

Kits

Many planners use a *kit system*. Parts required for an upcoming project or job are collected in what is called a *kit* and placed in a special bin or location. These parts are formally withdrawn from the storeroom, but may still be held somewhere else in the storeroom. The planner is assured that the parts will not be removed from the kit unless authorized by the planner.

Kitting is helpful to planners but can pose some problems to the storekeeper. Often the kit is developed for a job or project that is scheduled for completion far down the road. The items placed in the kit may be needed for other jobs that occur in the meantime. The storekeeper may turn away someone because they have no knowledge that the parts are sitting in a kit somewhere else in the facility. Another problem exists if the items removed for a kit bring the stores inventory to the reorder point. The storekeeper may reorder the parts even though they are still in the facility somewhere. An overstock condition can develop if the kitted job does not go as planned and the parts are returned to stores.

An improved system has been developed in many CMMSs. Parts needed for future jobs can be *allocated to the job* but left in the storeroom. This is accomplished in some CMMSs by making an entry in a field marked allocated or reserved. This system also has its problems. If someone comes to a storeroom and asks for a part that is familiar to the storekeepers, they may not check the computer record; they may just go and get the part. The breach into the reserved stock is not recognized until the withdrawal is entered into the stores program. This situation may be infrequent but it can create problems for the planner.

Tagging Parts

If an allocation field is not available, the parts can be marked on the shelf with a tag indicating the work order. Projected date to be used and the name of the person removing the parts are also placed in the spaces provided on the tag, as shown in Figure 2-17.

Any activity on the allocated parts must be reported to the individual who allocated the parts. If the parts are not used by the date indicated, the storekeeper can issue them freely.

NOTE

Allocation scheduling methods are discussed in detail in Chapter 5.

Saving Space

High inventory costs or space limitations may dictate that certain items cannot be held in the storeroom. Rather than letting the storekeeper make decisions in an information void, the planner/scheduler should work with them to reduce inventory. One option used at some facilities involves keeping parts *on the stores record* but *not in the storeroom*. The items commonly used are entered into the stores management system with *zero reorder quantity and no reorder point*. The part is identified with a stock number, vendor, pricing, and all other stores information, except a location in the storeroom. The storekeeper can order theses items as needed. The planner should investigate the following:

Figure 2-17   Part tag.

- Review equipment spare parts. Items may be held in the storeroom to rebuild equipment that is now being sent out for repair, or there may be sufficient complete spares, which means that the rebuild parts can be reordered as needed.

- Identify items that may have a short delivery lead time and that won't be needed very quickly after an equipment breakdown occurs.

- Identify items that are not needed for emergency repair.

- Identify items that are used only for preventive maintenance. The planner can notify the storekeeper when the preventive maintenance job is due to be performed and the storekeeper can then order the items required to perform the job.

- Help the storekeeper purge spare items for equipment that has long been removed or replaced with a different model.

Ordering Parts

Most of the items normally consumed in the performance of repeat jobs should be stocked in the storeroom. Since the usage of these parts is more predictable than parts used for emergency work, the storeroom may be able to *reduce the safety stock*. Safety stock is the amount of stock over an average reorder point used to account for delays in delivery or variable usage. The planner/scheduler should give the storeroom a list of the parts used for PM work. A schedule of the usage in the coming year would also be helpful to the storekeeper.

Special Orders

Under normal stores operations, specific items experience an average turnover. However, changes in usage can occur. The planner/scheduler should *notify the storeroom of the potential for increased usage* prior to a shutdown, outage, or major job.

Job Site Deliveries

This is a viable option for some larger maintenance organizations. Deliveries tend to cut down on travel time by maintenance workers, effectively reducing the time on any specific job.

Handling Delays in an Estimate

The normal workday for a maintenance worker contains delays. Planners are always looking for a consistent way to take delay into account when they develop an estimate. The solution lies in an understanding of the source of the work delay by the planner.

Figure 2-18 shows how *some delay* is built into the workday. The day in this example is normally 8.5 hours long, starting at 8:00 A.M.

The Workday

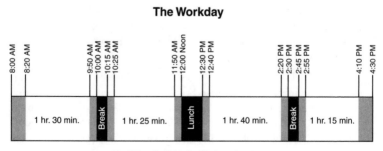

Total Time Available for Work = 5 hr. 50 min.

Figure 2-18 The workday.

and ending at 4:30 P.M. An unpaid lunch is taken from noon to 12:30 P.M., making the workday 8 hours long. Two 15-minute paid breaks are provided in the morning and the afternoon. If these company-sanctioned breaks were the only delays a planner had to account for, the job of estimating would be greatly simplified. Unfortunately, there are a number of other delays in the workday that occur with management's tacit approval.

Delays really begin first thing in the morning. The workday usually starts with the supervisor handing out jobs to the employees and giving some instructions about these jobs. Not everyone can be given instructions at the same time. The majority of the work force will be delayed *waiting for instructions,* while others will be delayed for a shorter period of time when they are actually receiving instructions. The maintenance worker will then have to spend some *preparation time* looking for tools and equipment that will be required on the job. This will always take some time out of the workday, even if the job is well-planned. *Traveling to the job* can also be considered part of the delay in the morning. The total morning delay in this example takes 20 minutes, on the average, out of the day (identified by the first shaded area).

Even though the break is at 10:00 A.M. in this example company, work usually stops at 9:50 A.M. The maintenance employees use the time before the break to *travel to the break area* and for some *personal cleanup* before they stop. The break is formally over at 10:15 A.M., but additional break time, called a *late start,* is usually taken in the break room and time is spent *traveling back to the job.* The time away from the job for the morning break has now grown to 35 minutes.

Lunch begins at noon but most work stops at 11:45 A.M. *Travel to the lunchroom* and *personal cleanup* take up most of this time. Work does not begin after the unpaid lunch period is over. Rather, additional lunchtime (or late start) is spent in the lunchroom and additional time is spent *traveling back to the job* site. The unpaid lunch ends up taking 50 minutes out of the paid workday.

The afternoon break goes much like the morning break and the 15-minute break time has effectively taken 35 minutes from the workday.

The day ends at 4:30 P.M. but most work stops at 4:10 P.M. The 20 minutes at the end of the day is spent on *traveling from the job site* to the shop, *putting tools away,* and on *personal cleanup.* Some companies provide a paid cleanup time at the end of the day, whereas others simply assume it will happen and don't bother enforcing the time.

This example assumes only one job is performed all day by each maintenance employee. This may seldom be the case and many jobs may be performed in a day. Each of these jobs will add to the travel time, as well as the time waiting for and receiving instructions.

The total time available for work during the 8-hour workday totals 5 hours and 50 minutes, or about 73 percent of the paid day. Some companies report an actual workday much less than this example, whereas others feel the above scenario does not fit their facilities. One way or another, some time is taken out of the workday for the following activities:

- Late starts
- Early quits
- Receiving instructions
- Waiting for instructions
- Travel time
- Paid breaks
- Personal cleanup time

Many companies attempt to measure, and then retrieve, the time lost through management improvements. Unfortunately, most of them report long-term failure. This lost time is uncorrectable, as far as the staff planner is concerned. The planner only needs to take this time into account when developing an estimate or putting together a schedule.

There are, of course, other delays in the remaining part of the workday. During the remaining 5 hours and 50 minutes in our example, the following events can occur:

- Correcting mistakes
- Waiting for parts in the storeroom
- Waiting for operations
- Waiting for another craft
- Rest breaks at the job site
- Lost motion
- Personal time
- General idle time (human nature)

All jobs include some or all of these delays. For the most part, it is unavoidable. Delay is particularly unavoidable *when two or more people work on a job*. An example is the replacement of a motor. In some cases this job may require two people to complete: an electrician and a mechanic.

The first part of the job requires the motor to be picked up and brought to the job site. Only one person (the mechanic or the electrician) is required for this step, so one may be idle. Both people must be involved in the lock-out of the equipment, but only the electrician is needed to electrically disconnect the motor. If sufficient space is available for both individuals, the mechanic can uncouple the motor from the driven equipment at the same time. At this point, the electrician's work is completed until the new motor must be reconnected. Depending on the size of the motor, the mechanic may need some help to position and move the new motor during the alignment process, but this requirement is limited and infrequent. It is unlikely the electrician can be *reassigned during the idle period* of this job. Temporary reassignment of maintenance workers often creates additional delay (such as travel time).

Most of the work performed by the electrician will be independent of the mechanic. The total time spent by the electrician actually working on the job will be much less than the mechanic. The *elapsed time* of the job will equal the time spent by the mechanic. The *labor hours charged to the job* will be two times the elapsed time. In reality the electrician was probably gainfully employed one-fourth of the elapsed time.

Keep in mind that the elapsed time of the job was less than it would have been if only one person (with electrical and mechanical talents) had been assigned to the job. The inefficiency associated with more than one person on the job may be tolerated if equipment *downtime must be kept to a minimum*. However, there is an upper limit to the number of people who can be assigned to a job and still reduce the downtime the job creates.

Table 2-11 shows how inefficiencies can be tolerated, to a point, if downtime must be limited.

Sixteen (16) labor hours are estimated for this job by the planner. If two pipefitters are assigned, the job should take 8 hours (16 labor hours divided by two pipefitters), resulting in 8 actual hours of downtime for the job. In this first case, the elapsed ideal

Table 2-11   Job: Replace Acid Line

Est. Labor Hrs.	Crew Size	Elapsed Ideal (Est./Crew)	Elapsed Actual	Actual Labor Charged
16 Labor Hrs.	2 PF	8 Hrs.	8 Hrs.	16 Labor Hrs.
16 Labor Hrs.	4 PF	4 Hrs.	4 Hrs.	16 Labor Hrs.
16 Labor Hrs.	6 PF	2.7 Hrs.	3 Hrs.	18 Labor Hrs.
16 Labor Hrs.	8 PF	2 Hrs.	3 Hrs.	24 Labor Hrs.

equals the actual elapsed time, because the two pipefitters can work independently and out of each other's way. The labor hours charged to the job should equal the estimate.

In an attempt to shorten downtime on the line, more people can be added to the job. Downtime is cut in half if four pipefitters are used on this job, because the ideal elapsed time again equals actual elapsed time. Again, only 16 labor hours are charged to the job.

Downtime should be cut to 2.7 hours if six pipefitters are on the job. Unfortunately, inefficiencies creep into the job because of *limited space* and *redundant activity*. The actual elapsed time is really 3 hours. The labor hours charged to the job have increased to 18 hours because of this inefficiency.

When eight pipefitters are assigned to the job, it still takes 3 hours to perform, so 24 labor hours are charged to the job. The extra pipefitters provide no increase in efficiency and no decrease in downtime. Six pipefitters should be assigned to the job if low downtime is important. Otherwise, two pipefitters will suffice.

Summary

All estimating processes have some form of inaccuracy. Good estimating involves good judgment. An estimate based on judgment is founded on the authoritative opinion of the planner. Estimates based on judgments are more likely to be met than those determined through guesswork. The accuracy of an estimate depends on the training and experience of the estimator and the quality of the data available to the planner. Data quality can be improved though investigation of the job and good data-collection techniques.

Planners, supervisors, and managers use estimates developed for maintenance work in various ways. Depending on the form, an estimate can be used to identify and control daily activities of the maintenance work force, or it can be used to control the cost of maintenance in a facility. Estimate types can be divided into two forms: hour estimates and dollar estimates.

All estimating methods require that a planner be trained in a specific way of thinking. The process should be logical and repeatable. The result of the process should be an estimate that closely approximates the actual steps and time required to complete a job. Common estimating methods used in industry today include construction planning and estimating, methods time measurement (MTM), the planning thought process, and estimates based on past performance.

MTM-based estimates do not take delay into account. Delay in MTM is considered to be the time it takes to do the job minus the standard time set by a benchmark. Additionally, an MTM-based process fails to identify needed materials, parts, tools, and equipment as other estimating methods do. The goal of a planning effort should be to provide an estimate, a list of tools, parts, and equipment required, but not a time standard to measure worker productivity.

A plan developed through the planning thought process differs greatly from one developed using MTM and other so-called engineered estimates. Steps developed through the planning thought process can be used by a maintenance worker as a guide through the completion of the job, if this is required. Additionally, planning thought process estimates are developed in more manageable, and some may say realistic, time units.

Estimates based on history or past performance can become the planner's most valuable resource in daily planning. The thoroughness and quality of such a resource is often hampered by inaccurate information that ends up in the historical record or important information that is excluded from the records. The best improvement in estimates comes from comparing job plans to the actual work-in-progress. As the work is being performed, a planner will compare what is actually being performed to what is happening in the field or on the floor.

Many planners collect all documents required to perform a job into a work package. A work package is a folder or group of folders that include all the necessary paperwork and reference materials required to complete a job.

Many factors affect the accuracy of the estimate, including planning conditions, direct planner problems, and insufficient time.

Material shortages and the failure of the planner to identify materials to be used on the job are major contributors to delays of work in progress. It is definitely important that the planner attain the parts. However this effort should not take most of the planner's time. Often, working with the storeroom is the best answer.

Planners are always looking for a consistent way to take delay into account when they develop an estimate. The solution lies in an understanding of the source of the work delay by the planner. Many companies attempt to measure, and then retrieve, the time lost through management improvements. Unfortunately, most of them report long-term failure. This lost time is uncorrectable, as far as the staff planner is concerned. The planner only needs to take this time into account when developing an estimate or putting together a schedule.

Chapter 3 provides some basic knowledge of electrical distribution systems and some planning tools while providing a real-world example of planning and estimating electrical work.

Chapter 3

Planning and Estimating Electrical Work

Now that we have a firm grasp of the fundamentals presented in Chapter 2 for planning and estimating, a logical next step is to apply some of that new-found knowledge. A solid real-world example of planning and estimating appears in the world of the electrician.

Planners with a strong background in mechanical and pipefitting trades usually feel at a loss when it comes to estimating electrical work. It is not always necessary to be an expert in the field to produce a usable electrical work plan, though. Electrical system design has been standardized through the years. The electrical field is probably the best-documented field of expertise. With some basic knowledge of electrical distribution systems and some *planning tools,* the novice planner can develop effective plans.

This chapter begins by providing some essential background needed to understand the needs and necessities of planning and estimating electrical work. Then we'll examine a real-world scenario involving an electrical installation project.

Background

The electrical power industry was developed through the efforts of some unique individuals around the turn of the century. Thomas Edison, one of the best-known inventors in the world, invented the electric light bulb in 1888. He then focused on the improvement of the direct-current (DC) generator and motor. DC-generating companies sprang up throughout the country. However, DC power distribution was limited to short runs because the DC-generating facilities had to be close to the end user to limit line losses. Edison dreamed of a day when small DC-generating facilities would be scattered around the country.

Alternating current (AC) power was available during this time, but was only usable for powering light bulbs. This was true until inventor Nikola Tesla, in partnership with businessman George Westinghouse, developed the AC induction motor. This invention changed the course set by Edison for the electrical industry.

Long runs of electrical cabling are possible with AC power. Line losses are limited by raising the voltage, reducing the current, and then reversing the process at the user's end. Lower current translates to lower line losses, a feat that is not possible with DC power.

Table 3-1   Common Transmission Voltages

Medium	High
2400 volts	69,000 volts
4160 volts	115,000 volts
12,470 volts	138,000 volts
13,200 volts	230,000 volts
13,800 volts	
24,940 volts	
34,500 volts	

This step-up and step-down process is accomplished using *transformers*. Transformers incorporate the magnetic induction principle to convert a low voltage at high current to a high voltage at a lower current, and vice versa. The voltage and current conversion is accomplished with marginal power loss.

Table 3-1 shows common transmission voltages found at the entrance of a facility's power system.

Most machinery is not designed to use more than 13,800 volts. To reduce the transmission voltage to a more usable level, step-down transformers are required at remote utility substations or at the end user.

Early AC power was generated at a *frequency* of 25 cycles per second, which is now called *hertz* (Hz). [The term was named for Heinrich Rudolf Hertz (1857–1894), who proved that electricity can be transmitted in electromagnetic waves, which led to the development of wireless telegraph and the radio.] Soon it became apparent that 50-Hz or 60-Hz AC power was the optimal distribution frequencies. Europe and the Eastern Hemisphere have standardized to 50 Hz. The Western Hemisphere has standardized to 60 Hz.

The AC induction motor also contributed to another modern standard. AC motors using more than 10 horsepower tend to vibrate less if they incorporate a *three-phase* design. Motor manufacturers promoted the use of three-phase motors, and three-phase distribution systems became the standard around the world.

Understanding the Single-Line Diagram

A three-phase AC power distribution system requires at least one cable for each phase. If a typical power system were sketched out with all cables and switches, the final drawing would be extremely complex and difficult to follow. Instead, a form of functional shorthand is employed. A single line is used to represent the three phases, and a three-phase device (such as a motor starter) is represented by a part associated with only one phase. This representation is called a *single-line diagram.* As shown in Figure 3-1, the single-line diagram is less complex than most electrical drawings and is easily understood by non-electrical personnel.

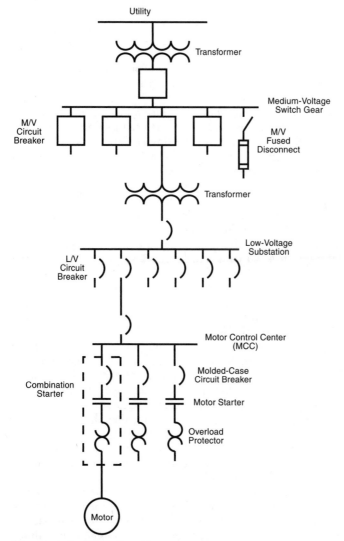

Figure 3-1 Single-line diagram.

Table 3-2 Common Distribution Voltages

Low	Medium
240 Volts	2400 Volts
480 Volts	4160 Volts
	12,470 Volts
	13,200 Volts
	13,800 Volts

Starting at the utility, the first component in a power system is the *transformer*. In most facilities, transformers can be *dry-type* or *oil-filled*. Dry-type transformers are usually restricted to medium voltage levels, whereas oil-types can be found in medium- or high-voltage systems. Common insulating oils used in transformers are mineral oil, silicon, or PCB (Polychlorinated Biphenyls). However, because of environmental concerns, PCB-filled transformers are used less and less. Some transformers rated for high-voltage operation may be *gas-filled*.

The high-voltage side of the transformer is called the *primary* and the low-voltage (distribution) side is called the *secondary*. Table 3-2 shows common distribution voltages.

Some large motors are designed to operate with a medium voltage supply of 2400 or 4160 volts. However, most motors are powered from a 480-volt secondary.

As shown in Figure 3-1, the horizontal line on the secondary side of the transformer is called a *bus*. Electrical buses can be long copper bars in a distribution system, one per phase. In general, a bus is any connection of three or more cables.

Feeding off the bus in the sample diagram are *medium-voltage distribution switch gears*. Medium-voltage circuit breakers are represented by a box. All *low- and medium-voltage fuses* and *fused disconnects* are represented by a vertical rectangle with two bars.

The *main breaker* is above the bus and the *distribution breakers* are below the bus. The top of each breaker is called the *line side,* and the bottom is called the *load side*. Medium-voltage breakers can be described as large, spring-loaded switches that can be tripped open by external relays (not shown). Medium-voltage breakers can be air-type, vacuum-type, or oil-filled.

Each of the distribution breakers feed through three-phase cables to another transformer. This transformer reduces medium voltages further to the more usable 240 or 480 volts. On the secondary side of this transformer is a *main low-voltage circuit breaker*. Low-voltage breakers are represented by a crescent. The main breaker feeds the bus that has feeder breakers, which is designed to protect the feeder cable on the load side. The main breaker and feeder breakers can be *metal clad breakers* with removable parts or *molded case breakers* that usually cannot be dismantled.

Each feeder breaker feeds a set of three cables (represented by one line). These cables usually terminate at a low-voltage *motor control center (MCC)*. The MCC is represented by one main breaker (usually molded case), feeding a bus that feeds other molded case breakers. The load side of each molded case breaker feeds a *motor starter* or *contactor*, represented by two parallel horizontal lines. Shown on the load side of the starter is the symbol for a *thermal overload protector*.

The molded case circuit breaker, contactor, and overload protective device are often packaged in the same enclosure called a *combination motor starter*. Cables connected to the load side of a motor starter feed to the end user of most electrical power, a *motor*. The symbol for a motor is a circle, either with the letter M or with the horsepower of the motor inside the circle.

The planner should get a copy of the single-line diagram for any electrical equipment serviced from the facility's engineering department or the electric shop. Walking down the system using the single-line diagram can help acquaint a novice with the system components just described. A few good books can also help you learn about an electrical distribution system, including the following:

* *Industrial Power Systems Handbook* by Donald Beeman (New York: McGraw-Hill Book Co., 1995)
* *American Electricians Handbook, Fourteenth Edition* by Terrell Croft and Wilford I. Summers (New York: McGraw-Hill Book Co., 2002)
* *Standard Handbook for Electrical Engineers, Fourteenth Edition* by Donald G. Fink and H. Wayne Beaty (New York: McGraw-Hill Book Co., 1995)
* *National Electrical Code* by the NGPA, (Quincy, MA: National Fire Protection Association, 2001)

Electrical estimating problems that a planner may confront can be separated into three categories:

* Estimating upgrades or new installations
* Estimating troubleshooting and repair jobs
* Identifying and estimating maintenance work during an electrical shutdown

Each of these categories is approached with a specific estimating method.

Many electrical departments are required to provide routine upgrades to the facility's electrical distribution system. This is less true for mature operations, but more true for office complexes, institutions, and research facilities. If a large part of electrical work is upgrade or new installation, standard tables should be developed by the planner to help build estimates.

The following sections provide a starting point for the planner. Adjustments should be made based on experience.

Branch Circuit Checklist

Branch circuits are wiring, conduit, and hardware for low-voltage equipment installations (such as lighting, heaters, and motors). The following checklist can be used to jog the planner's memory when installing new equipment or applications:

- *Conduit*
    - Conduit (rigid or thin wall)
    - Greenfield
    - Elbows
    - Couplings
    - Greenfield connectors
    - Thin wall couplings and connectors
    - Locknuts and bushings
    - Straps
    - Hangers
    - Anchors and screws
- *Outlet Boxes*
    - All types and sizes
    - Plaster rings
    - Covers
    - Extensions
    - Fixture studs
    - Hangers
    - Toggle bolts
    - Expansion shields
    - Special outlets
    - Condulets or unilets

- *Wire*
  - Wire type (T, TW, RW, or RH)
  - Wire connectors (wire nuts, lugs, and so on)
  - Tape
  - Solder
- *Wiring Devices*
  - One-, three-, four-way switches
  - Duplex receptacles
  - Grounding receptacles
  - Clock receptacles
  - Special receptacles
  - Plates
  - Plates (special finish)
  - Special devices
  - Weatherproof switches and receptacles

Feeder Checklist

Feeders are the cables leading to lighting panels, switchboards, and fuse boxes. Following is a list of components required to install feeders:

- Conduit
- Wire
- Cable
- Pull boxes
- Insulated bushings
- Locknuts and bushings
- Lugs and tee taps
- Tape
- Insulating materials
- Steel rod
- Angles
- Channels
- Pipe straps and clamps
- Iron fish wire
- Bolts

- Beam clamps
- Hole cutting and drilling tools
- Elbows
- Conduit fittings
- Grounding
- Expansion shields
- Concrete inserts
- Sleeves
- Cable supports
- Hangers
- Couplings
- Expansion fittings
- Unions
- Welding and cutting equipment

Conduit Installation Time

Table 3-3 provides some standard guidelines for estimating rigid conduit installations. An allowance is included in the figures for an average number of offsets and bends.

Pulling Cables in Conduit

Table 3-4 provides the time required to pull a cable through a conduit.

For example, pulling a 300 MCM through 250 feet of conduit requires the following time:

300 MCM cable will take 45 hours per 1000 feet to install (derived from Table 3-4),

45 hours/1000 feet × 250 feet = 11.25 hours.

Table 3-3   Guidelines for Estimating Rigid Conduit Installation

Conduit Sizes	Hr. per 100 ft
$\frac{1}{2}''$ to 1″	4
$1\frac{1}{4}''$ to $2\frac{1}{2}''$	8
3″ to 4″	16

For open, clear space construction and for ceilings 12 feet or less in height.
Additions: *Add 20 percent where ceiling height is 14–16 feet; add 40 percent where ceiling height is 18–20 feet.*
Deductions: *For two or more conduits in same run, multiply estimate by number of individual runs and deduct 10 percent from total estimate.*

Table 3-4   Time Required to Pull Cable Through a Conduit (Hours per 1000 Feet)

Size of Wire	100 ft or less	101 to 200 ft	201 to 300 ft	More than 300 ft
4/0	44	40	37	31
250 MCM	49	45	41	35
300 MCM	56	50	45	38
350 MCM	60	55	49	39
400 MCM	65	60	54	42
450 MCM	75	65	58	44
500 MCM	80	70	65	48
600 MCM	95	80	72	55
700 MCM	100	88	85	63
750 MCM	110	94	88	68
800 MCM	112	98	90	72
900 MCM	122	108	98	82
1000 MCM	130	120	108	90

Mounting Lighting Panels

Table 3-5 can be use to estimate the time required to install a lighting panel and make all necessary connections.

Motor and Motor Control Checklist

The following checklist may be helpful when planning motor installations:

- Anchors
- Belts

Table 3-5   Estimating Time Required to Install a Lighting Panel

Branch Circuits (30-amp, 2-wire)	Flush Mounting Hr./ Panel
5	11
10	14
15	18
20	20
25	24
30	28

Note: Deduct 10 percent for surface cabinets.

- Bolts
- Breakers
- Coupling
- Disconnect switches
- Flexible connections
- Fuses
- Grounding jumpers
- I.D. plates
- Lugs
- Motor starters
- Motors
- Mounting and fastening materials
- Pilot lights
- Pushbutton stations
- Safe-stop stations
- Sheaves
- Tape

Time Estimates for Mounting a Motor

Table 3-6 shows the time required to mount a motor on a base that is prepared for the motor (pre-existing anchor studs or bolt holes). The estimate shown includes sheave or coupling alignment.

Estimating Time to Connect a Motor

Table 3-7 shows time estimates to connect up a motor, including the installation of flexible connections and checking for rotation.

Table 3-6   Time Required to Mount a Motor on a Base

Motor Size	Hours per hp
Up to 3	1.0
5 to 10	0.25
15 to 40	0.35
50 to 100	0.5

Table 3-7 Time Required to Connect a Motor (Labor Hours)

Size	AC	DC
Up to 3 hp	0.25	0.50
5 to 10 hp	0.5	0.75
15 to 25 hp	0.75	1.0
30 to 50 hp	1.0	1.5
60 to 125 hp	1.5	2.5

Medium Voltage (601 to 35,000 Volts) Feeder Checklist

The following list can be used by planners to help the determine cable requirements:

- Arc proofing
- Cable
- Cable support brackets
- Cable tap boxes
- Conduit cable supports
- Ground wire
- High potential test
- Insulating oil
- Insulating putty
- Lead
- Lugs and connectors
- Potheads
- Pulling compound
- Shielding braid
- Special equipment rental
- Splice kit
- Tape, all types
- Termination materials

Estimating Troubleshooting and Repair Jobs

People who plan electrical work often complain that the actual time spent on a job can vary widely from the estimated hours. The planners claim there is no way to know the problems that may be

Table 3-8   Work Completed on Lights-Out Problem

Cause	% Occurrence	Average Time to Fix
Burnt-out light bulb	87% of the time	15 minutes
Tripped circuit breaker	9% of the time	21 minutes
Other problem	4% of the time	1.3 hours
	Average Job	18 minutes

uncovered during the troubleshooting process. For example, a simple work request stating the lights are out may require only 15 minutes to replace a bulb, or it could take two days to replace burnt-out wiring feeding the light bulb. This lack of predictability is often used as a reason for never planning electrical work.

In reality, most nonemergency troubleshooting jobs fall into a narrow range of predictability. Returning to our lights-out example, a review of completed work at a chemical facility resulted in the data shown in Table 3-8. As shown here, about 95 percent of the time this problem only took about 20 minutes to correct.

Following are three ways to approach plans for troubleshooting or repair jobs:

- Estimate for the worst case.
- Estimate for the average job.
- Estimate for the most common case.

An estimate based on the *worst case* in our lights-out example would allot an extra hour 95 percent of the time to lights-out jobs the electrician is scheduled to perform. The estimate is inflated 400 percent most of the time. The schedule may be completed with at least an hour to spare if this were the only job on the schedule that was estimated for the worst-case scenario. However, if all jobs on the schedule were estimated for the worst case, the schedule could be inflated to 400 percent or more on any given day. Worst-case estimating does not provide a supervisor with a workable and realistic schedule.

Estimates based on the *average job* cause the same scheduling problem as worst-case estimating, but not as pronounced. In our lights-out example, the average job took 18 minutes. This means that most of the time the estimate was inflated to 120 percent. In a normal workday, the schedule will be inflated by about an hour and a half most of the time.

The average job estimate is often used by electrical contractors rather than in-house planners. Electrical contractors try to meet the need of the client when bidding on a job but must also protect themselves from unforeseen circumstances (which could cause a loss of profit). Facility planners are not faced with a loss of profit.

The *most-common case* estimate works best for many planners. A schedule developed using most-common case estimates may not be finished on some days. However, this is not usually a problem because the schedule is not always finished because of emergencies and other events that break the schedule. The planner should use only most-common case estimates and list the jobs in order of importance for the supervisor.

For those who feel that a plan is not necessary for electrical work, it is important to note one other benefit of planning aside from a consistent schedule. Developing a plan for electrical maintenance helps develop a *parts list* and *tool list*. A list of parts and tools will always help an electrician be more productive.

Identifying and Estimating Maintenance Work During an Electrical Shutdown

Sometimes when planning for a shutdown or turnaround in industrial facilities, an electrical outage may be required. Planning for this outage often is overlooked. However, proper planning for this electrical outage can provide an opportunity to define future required work.

Appendix A provides a partial list of work before and during an electrical outage, as well as a sample checklist for planning the shutdown.

Summary

The electrical power industry was developed through the efforts of some unique individuals around the turn of the century. Inventor Nikola Tesla, in partnership with businessman George Westinghouse, developed the alternating current (AC) induction motor, which changed the course set by Thomas Edison for the electrical industry with his direct current (DC) generator and motor. Long runs of electrical cabling are possible with AC power. Line losses are limited by raising the voltage and reducing the current and then reversing the process at the user's end. Lower current translates to lower line losses, a feat that is not possible with DC power.

This step-up and step-down process is accomplished using transformers. Transformers incorporate the magnetic induction principle to convert a low voltage at high current to a high voltage at a lower current, and vice versa. The voltage and current conversion is accomplished with marginal power loss.

The AC induction motor also contributed to another modern standard. AC motors using more than 10 horsepower tend to vibrate less if they incorporate a three-phase design. Motor manufacturers promoted the use of three-phase motors, and three-phase distribution systems became the standard around the world.

A three-phase AC power distribution system requires at least one cable for each phase. If a typical power system was sketched out with all cables and switches, the final drawing would be extremely complex and difficult to follow. Instead, a form of functional shorthand is employed. A single line is used to represent the three phases, and a three-phase device (such as a motor starter) is represented by a part associated with only one phase. This representation is called a single-line diagram. The planner should get a copy of the single-line diagram from the facility's engineering department or the electric shop. Walking down the system using the single-line diagram can help acquaint a novice with the system components.

Electrical estimating problems that a planner may confront can be separated into three categories.

- Estimating upgrades or new installations
- Estimating troubleshooting and repair jobs
- Identifying and estimating maintenance work during an electrical shutdown.

Each of these categories is approached with a specific estimating method.

Chapter 4 introduces the concepts of preventive and predictive maintenance and discusses some of the details surrounding policies and procedures related to each.

Chapter 4

Understanding Preventive and Predictive Maintenance

Two of the most effective programs used to cut costs related to equipment failure and plant downtime are preventive maintenance and predictive maintenance.

Preventive maintenance is basic maintenance performed on machinery or facilities at an established interval or frequency. The main purpose of PM is to extend the equipment life and ensure capacity. PM can also be applied to protect personnel or the environment. PM can be performed when the equipment is shut down, or it can include adjustments made while the equipment is running. PM work can be anything from a simple meter reading to a major rebuild of equipment.

The *frequency* with which PM is performed is determined by the maximum period of time allowable before breakdown or extensive corrective work is expected. The establishment of this interval is vital in limiting the cost of PM without sacrificing the value of the effort. The frequency is usually set at a test value. This value can be a short interval that may, in time, prove to be too frequent. The cost of this PM routine will usually exceed its return until the duration between PMs is decreased. The frequency can be based on manufacturers' data or historical data. Adjustments may be required to these frequencies, as well, if a failure occurs between scheduled PM.

One improvement on PM is *predictive maintenance* (PDM). First considered as an adjunct to most PM programs, PDM is now recognized as distinct from PM efforts. Though PM tasks are initiated by a calendar (based on a set frequency), PDM programs measure and react to the condition of operating equipment.

PDM compares the trend of measured physical parameters against known engineering limits for the purpose of detecting, analyzing, and correcting problems before failure occurs. A predictive approach can be applied if first a physical parameter (such as vibration, temperature, pressure, voltage, current, or resistance) can be measured. An engineering limit for the measured physical parameter must be established so that a problem can be detected during routine monitoring. Hopefully, the limit is low enough to detect the problem before excessive damage occurs. The repair process can be enhanced with a good analysis of the problem. The correction of the root problem is the key to most predictive efforts.

This chapter takes a detailed look at both PM and PDM programs. The discussion begins with an examination of the relationship between PM costs versus production costs.

Examining Preventive Maintenance (PM)

Every discussion of preventive maintenance (PM) includes the classic curves that relate the costs associated with PM and the costs associated with production. These curves are often used to substantiate the benefits of PM, as shown in Figure 4-1.

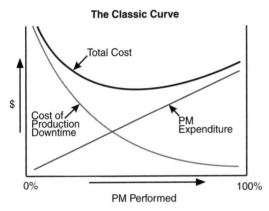

Figure 4-1 The classic curve.

The relationship shows that costs attributed to operations downtime are high when expenditure on PM is low, resulting in a very high total cost. As more money is spent on PM, costs attributed to lost production, as well as the total cost, go down until a minimum cost is achieved.

The assumption made by this curve is that the PM dollars are being spent in the most effective manner on valid and useful procedures. Everyone recognizes the well-worn benefits of PM, such as:

- *Reduced cost* of maintenance
- *Increased on-line time* of operations equipment
- *Higher worker productivity* while performing the PM

However, prior to beginning a full-fledged PM program, people commonly worried that they *might perform too much PM work* and derive little value from each additional PM dollar. This is the point on the curve at which PM expenditures are high, but the cost of production downtime has been reduced greatly. The fact is that these people are usually nearer the low PM expenditure–high production downtime, part of the curve. The time to start worrying about performing too much PM is when the low point on the total cost curve is reached.

No one should be deterred from beginning or expanding a PM program for fear of spending too much money because other factors tend to *invalidate the curve*. Modern market requirements insist that production operate at an optimum level to maintain market share. The higher total costs for operating near the right-hand side of the curve become acceptable to avoid lost production. Lost production can mean losing a customer in a tight market.

Because of the more extensive and costly damage possible from equipment failure and the need for longer intervals between shutdowns, *thorough PM* procedures are required. PM activities can be broken into two basic categories:

- Mandatory
- Discretionary

The first step in any maintenance effort is to meet *mandatory PM requirements*. Mandatory PMs are activities performed on equipment or facilities as required by law or contract. Federal agencies such as the Occupational Safety and Health Organization (OSHA), Environmental Protection Agency (EPA), Department of Transportation (DOT), Nuclear Regulatory Commission (NRC), or Federal Energy Regulatory Commission (FERC) require inspections and tests to protect facility personnel and the general citizenry. State and other municipal agencies have inspection requirements for pressure vessels and boilers. Insurance policies on large equipment, or on an entire facility, usually mandate rudimentary PM measures. Manufacturer warrantees may also stipulate periodic maintenance or inspections. These requirements should be treated as mandatory if a company hopes to attain the full benefit. Mandatory PM procedures usually require documentation to verify that the work has been performed. Failure to provide such documentation can result in fines (in the case of government-mandated PM) or might invalidate operating licenses and warrantees.

Apart from mandatory requirements, the bulk of a well-rounded PM effort would include discretionary work. *Discretionary PM* activities performed on equipment or at a facility are based on the need to reduce downtime, minimize costly equipment damage, or ensure personnel safety. Although all equipment can be considered a candidate for PM, a process of prioritization and a cost-versus-benefit analysis is usually performed to justify discretionary PM activity.

An *ABC analysis* looks at the cost to maintain equipment or the frequency of failures. In such an analysis, equipment maintenance history and repair costs are reviewed. Typically, 20 percent of the machines in a process account for approximately 80 percent of the maintenance and downtime costs. These machines, called the A items, are prime subjects for PM consideration. The B and C items are the remaining 80 percent of the machines that account for only 20 percent of the maintenance and downtime costs. Once the A items are adequately covered by a PM program, the B and C items can be added based on a cost-versus-benefit analysis.

Another approach to prioritizing candidates for discretionary PM would be to conduct a *vulnerability study*. All equipment in a facility is classified according to its relative effect on the operation or on its potential for reducing revenue. A committee of select personnel reviews the entire list of equipment in a facility. Questions such as the following are asked about each piece of equipment or assembly:

- What is the average cost of downtime for this assembly?
- What is the replacement cost of this assembly?
- What percentage of production would be lost if this assembly failed?

A weight of one to ten is then placed on each item, based on the relative cost to the company. The top 20 percent of the equipment on the list are then chosen for an initial effort. This list may be extended once the cost of additional PM work is compared to the cost of lost operation.

Lubrication is one type of discretionary PM that most people find to be essential, even without a cost-versus-benefit analysis. A lubrication program is necessary to avoid premature failure of machinery and is usually inexpensive to perform.

Writing a PM Procedure

Once a list of equipment requiring mandatory or discretionary PM has been defined, a solid PM procedure must be developed to combat the root cause of the problem. All too often, a PM work order is handed to a maintenance worker without any more detail than a statement such as, "PM the plant air compressor," or "Inspect the gearbox." The mechanic is expected to draw on personal knowledge of equipment and to do all the preventive work required. What is actually accomplished can vary widely from one employee to the next, and the work performed may not even be the work intended by the manager. The time taken out of the normal workday to perform PM should be productive, so the work must be well-defined. A written *procedure* with a *data sheet* should be developed for most PM work.

The following types of PM procedures should be considered:

- Inspections
- Adjustments
- Testing
- Calibrations
- Rebuilds
- Replacements

Inspections

An *inspection* is usually performed on operating machinery using the basic senses of sight, sound, and touch. Unless specific parameters have been identified, the quality of the inspection is totally dependent on the sensitivity of the inspector to potential equipment problems. Associated checklists used on inspections are often just a list of equipment, the symptomatic conditions to be looked for, and a location on the checklist where a check mark is to be entered. Such checklists often become abused, with inspectors hurriedly filling in the check marks or even completing the forms back in the shop.

Inspections can be enhanced by adding gauges or meters to the equipment being inspected. Rotating components (such as couplings or belt drives) can be inspected under the freezing action of a battery-powered strobe light. Belt drive inspection can be expanded to include the use of sheave gauges and belt tension gauges. Critical temperature limits on bearings or compressor valves can be determined by using an inexpensive temperature infrared pyrometer. Vibration checks using low-cost, pen-sized meters can be used to detect beginning problems in equipment caused by imbalance, misalignment, or looseness. Adding such innovations to an inspection often raises the interest level of the inspector, as well as heightens the quality of information returned.

Adjustments

Adjustments involve the optimization of operating equipment. This could be a simple fine-tuning of a cam on a limit switch, adjusting the thrust of a rotary kiln, or could involve tuning a boiler combustion control system to maximize fuel efficiency.

Testing

Tests are used to verify that equipment is performing according to specifications. These are often associated with safety controls or environmental equipment. A typical PM of this type might be used to verify the automatic shutdown controls on a boiler steam drum. Documentation is often required when performing these types of tests. They should include a form that shows the date and results of the test, and should be signed by the tester.

Calibrations

A *calibration* is performed to verify or correct the accuracy of critical indicators, control instruments, or final elements. Rather than preventing downtime, calibrations ensure the in-spec operation of a process. Some calibrations are mandated by a permit to ensure environmental compliance. A data sheet should include a calibration procedure, which indicates an as found, correct, and as left condition. These sheets should be signed and dated.

Rebuilds

A *rebuild* invariably requires equipment downtime. During a rebuild, critical dimensions are checked and worn parts replaced. The goal of a rebuild is to restore equipment to like-new condition and to avoid production downtime or off-hours failure. Rebuilds are sometimes turned over to an outside vendor. The vendor may be the original equipment manufacturer (OEM) or a service vendor familiar with the machinery's maintenance requirements. A detailed report covering what was found, what work was done, and what should be checked on the next rebuild (at the conclusion of the rebuild), should be included in the PM record.

Replacements

Replacement PMs can involve the periodic replacement of disposable components (such as filters or lamps). Replacement PMs can also be a periodic change-out of equipment or components in anticipation of failure. This type of change-out is performed as insurance against an off-hours failure or failure during a production period. The replacement of an inexpensive diesel engine fuel oil pump at the beginning of the winter season is an example.

Procedure Elements

Good PM procedures should try to incorporate the following elements:

- A list of *tools, parts,* and *instruments* required to perform the PM should be provided at the beginning of every procedure.
- If the PM procedure includes taking *measurements* or *readings,* a form must be provided so that this information can be recorded. Typical information that should be recorded includes pressure and temperature levels, amps, volts, and pH.
- If measurements or readings are taken, the data form must include a *limit* or *range* of values that will indicate whether the measurement or reading is normal. The data sheet should also inform the PM mechanic what to do (if anything) when the reading or measurement falls outside the limit or range.
- The data form must ensure that measurements were actually taken, or that the technician had been to the site. This can be hour-meter readings, mileage, or trip counts on circuit breakers. Signed or punched tags, stickers, or labels on the equipment can also perform this function.
- *Safety considerations* should be listed (such as lock-out or hot-work procedures). Have the technician contact someone from operations in the area before beginning the procedure.
- If an *equipment shutdown* is required to perform the PM, be sure that it is arranged in advance. Don't force the maintenance worker to coordinate the job or to do PM routines that are ill-planned.

Encourage additional comments from the technicians performing PM work. Reacting to their comments and giving them visibility for their efforts will improve their involvement in the program.

PM Development Worksheet
The effort of generating PM procedures should be made as easy as possible. Persons unaccustomed to developing PM procedures need a worksheet to aid the process. Figure 4-2 shows one such worksheet.

This form is used to describe the following:

- The *area* in the facility in which the technician can find the equipment (such as a cost center).
- The name and number of the *equipment* to be maintained.
- *Manpower requirements* and estimated hours to perform.
- *Shutdown* or *preparation requirements* of operations.

PM Work Sheet

Equip. #	Equip. Name	CC	Prepared by:	Date

PM Title:

Craft: ______ Crew: ______ Estimated Crafthours: ______ Operations Requirements: ______

Reschedule Data:

Frequency: Daily Weekly Bi-monthly Monthly Semi-annual Annual Other ______
(Circle one)

Active Window: __/__/XX to __/__/XX Start Date: __/__/__

Reschedule From:
(Check one)

☐ Next PM Date Plus Freq.

☐ Completed Date Plus Freq.

Procedure:

Material:

Stock #	Quant.	Description

Tools & Equipment:

Safety Precautions:

Related Drawings: ______

Figure 4-2 PM worksheet.

- *Reschedule information,* including frequency, active window (such as summer-only for air conditioners), start date (first date to start the PM), method of recalculating the next PM date.
- Detailed *description* of work to be performed (on another sheet, if necessary).
- *Parts* and *materials* required to perform the work, including the storeroom stock number.
- *Tools* and *equipment* required to perform the work.

A Preventive Maintenance Example
Checks for Three-Phase Power Factor Capacitor Banks

Frequency: Quarterly

Tools Required: Amprobe · H/V Voltmeter or Hot-stick
Digital Multimeter · Electricians Tool Pouch
H/V Ampclamp

Safety Requirements:
Be sure not to use low-voltage test equipment on high-voltage lines. Occasionally, a lock out of the equipment will be necessary when changing fuses.

Procedure:

1. Check amps on all three phases and record the readings.

2. The amp readings should be within + or − 1% of each other. If not, circle the farthest one out.

3. The amp readings must agree within + or − 5% of the value calculated from the following equation:

$$\text{Amps} = \frac{\text{KVAR Rating} \times 577}{\text{Volts (Phase to Phase)}}$$

4. If there is a discrepancy in the readings due either to uneven or out-of-range amps, first check the voltage to the fuses, then determine if the fuses are blown. If one fuse is blown, disconnect the power to the capacitor bank and replace all three fuses. Reenergize the bank and repeat Steps 1 through 3. Note any changes made.

5. If the fuses blow again or a discrepancy in the readings still exists disconnect power from the capacitor bank and discharge the capacitors. Using a ohmmeter (not a megohmeter), measure phase to ground resistance on each phase of the capacitor bank. Any reading near zero ohms will identify the bad capacitor.
   Any bad reading on a "3 phase capacitor" means the whole capacitor needs to be replaced. If a discrepancy still exists discharge the capacitors again and isolate the bad bank from the rest of the capacitor set. Note the condition of the capacitors.

6. Retest the amps on each phase; all should agree as stated in 1 and 2.

7. If the fuses blow again, do not repeat the procedure. Disconnect power to the bank and report the condition.

Figure 4-3 PM example.

- *Safety* precautions.
- Related drawings, manuals, or checklists.

Figure 4-3 and Figure 4-4 show a PM example and an accompanying worksheet.

Sources of PM Procedures

The following sources will help identify PM procedures:

- Vendor-recommended PM
- Plant experience

Capacitor Bank Check Route List
    Name TM   **Date** 8/6/2003

Location	Notes	KVAR@ Volts	Amp Check			Comments
			A	B	C	
Sub #1 L/V Bank	217 TO 265 amps per ph	200 @ 480 v	250	250	250	
Sub #2 L/V Bank	217 TO 265 amps per ph	200 @ 480 v	250	250	250	
Sub #2 H/V Bank	Do not disconnect 54 to 66 amp	50 @ 4160v	0 / 60	0 / 60	0 / 60	Fuses Blown Replaced all Three
Sub #3 L/V Bank	108 to 132a per phase	100 @ 480v	125	125	125	
MCC 12	Two Banks. 81 to 99 amp per phase	75 ea. @ 480v	95	95	95	
MCC 6	Three Banks 81 to 99 amp per phase	75 ea. @ 480v	85	85	85	
Sub #4 L/V Bank	217 TO 265 amps per ph	200 @ 480v	270	270	270	Voltage is high (510v/phase)
Sub #5 L/V Bank	217 TO 265 amps per ph	200 @ 480v	240	240	240	

Figure 4-4 PM example worksheet.

- Generic PM
- Equipment history

Vendor-Recommended PM
The maintenance manuals provided by equipment manufacturers are invaluable when trying to build a PM procedure. However, the manual should not be used in its original form. Additional tests, repairs, or inspections should be added that reflect the special use of the machine in your facility. As an example, if a compressor normally operates at an overload, routine inspections should be more frequent than suggested by the manufacturer. Also, it may be advisable to perform tests on the lube oil to detect breakdown caused by high temperatures. You may want to include oil tests and other chemical analyses not usually suggested by vendors in their standard maintenance procedures.

Plant Experience
Mechanics who work on equipment and the operators that run it can help define PM procedures specific to the equipment's environment.

Generic PM
Procedures developed for general classes of equipment can be modified to fit a specific facility requirement. Table 4-1 shows some

Table 4-I   Key Contacts for PM Procedures

Organization	Contact	Standards
Institute of Electrical and Electronic Engineers (IEEE)	Publications Center, 445 Hose Lane, Piscataway, NJ, 08855, (800) 678-4333	Std. 43: Testing & Inspection of Large Motors Std. 190: Testing & Inspection of Small Transformers Std. 446: Emergency and Standby Power Std. 141: Electrical Power Distribution
Instrument Society of America (ISA)	Alexander Drive, Research Triangle Park, NC 27709, (919) 549-8411	Instrument Air Quality Measurement Instrument Maintenance Managers' Source Book RP31.11: Calibration of Turbine Flow Meters
American Society of Mechanical Engineers (ASME)	East 47th Street, New York, NY, 10017, (212) 605-3333	Pressure Vessel Testing Standards Pressure Relief Valve Tests and Rebuilds
National Fire Protection Association (NFPA)	Batterymarch Park, Quincy, MA, 02269, (800) 344-3555	Std. 70: National Electrical Code Std. 70B: Preventive Maintenance of Electrical Distribution Equipment

typical procedures of this type that were developed by professional organizations.

Equipment History

The maintenance history of equipment can also identify PM needs. The frequent breakdown of a particular machine might be prevented through some type of inspection, adjustment, or replacement of a component. History will aid in identifying an optimum frequency of such PM.

The quality of historical data is directly affected by the original diagnosis and repair information included in the history records. Often, the real problem is not identifiable through history information because of insufficient data noted in the nature of repairs. Technician's comments will often lead an astute analyst of maintenance history to a root problem that had not been addressed in the past.

Evaluating PM Procedures

All PM procedures must be regularly evaluated. The following checks must be made:

- *Is the procedure still valid?*—If the machinery has been replaced with another type, are the job steps the same, or must the procedure be changed?

- *Are the limits or ranges still valid?*—Process changes or changes in load or environment can change the normal operation range or limit. Limits or ranges are often set too conservatively. A good question to ask when validating ranges or limits is, at what point will damage be likely to occur?

- *Is the frequency still valid?*—Has a failure occurred between scheduled PM and could the failure have been averted through more frequent PM? Can some PM frequencies be stretched? Have the operating hours of the equipment been changed?

- *Can portions of the PM effort be moved back to operations?*— If so, what kind of feedback must be set up to inform maintenance of corrective work identified through operations surveillance activities?

Evaluating PM Programs

The PM program itself must be evaluated on occasion. You should consider the following:

- Both the open and closed work order files should be inspected for corrective work identified by the PM inspections. *As a rule, an effective PM program will generate three corrective action work orders or on-the-spot repairs for every 10 PM procedures performed.*

- Verify the last PM performed by visiting the site where the PM was performed and by reviewing the procedure.

- Verify that completed PM data sheets are filled out properly and that action is being taken when measurements or readings are out of range.

- Overdue PM work orders should be identified and the reason they were not scheduled should be determined. The frequency of the PM may require a PM procedure to be performed too often, in which case, the frequency should be lengthened. If on the other hand, the importance of the PM is not understood by those responsible for getting it on the schedule, more visibility should be given to the consequences of not completing the PM procedures.

- The amount of and reasons for equipment downtime is the ultimate evaluation of the PM effort. Tabulation of this parameter provides the program with visibility, whether it is positive or negative.

Unrecognized Benefits of PM

Everyone should recognize the time-worn benefits of PM, but often left unrecognized are the hidden benefits for maintenance. Those benefits include the following:

- PM should determine a major portion of the work that maintenance performs on a day-to-day, week-to-week basis. Because this work is well-defined, the work is preplanned as far as labor estimates, materials, and other needed resources. Because it carries a high priority, the personnel required to perform it are already allocated. Thus, a major portion of the maintenance planner's work is already completed, and the planner can concentrate on developing solid work plans for the remaining corrective-type work orders.

- Because the supplies and materials required for PM are well-defined, those materials can be purchased just prior to the scheduling of the work, minimizing inventory dollar values.

- A well thought-out and implemented PM program can dramatically reduce emergencies. The end result is that even corrective work can be completed without frequent stops. Besides adding to the overall productivity of the maintenance effort, the working climate of maintenance is improved and the morale of the work force is boosted.

- An ongoing PM effort on major machinery helps identify the major overhaul work that may be required. Maintenance is not surprised by the problems encountered when the machinery is taken down for such work.

- The parts or materials usage in the facility becomes more predictable when a PM program is in force. Large insurance stock will become unnecessary. The minimum inventory levels of these items can be attained.

- The budgeting process is streamlined when a PM program is in place, since much of the work that maintenance is performing is already identified. The manager knows what the labor requirements, necessary materials and supplies, and major rebuilds that must be planned are.

Other Options and Pitfalls to Avoid

A PM procedure should be developed to record hour-meter or count-meter readings. Running time limits may elapse for equipment if no one is monitoring the meter readings. This type of work is often given to operations personnel to record, but maintenance personnel may never see the log sheets.

Be specific as to what is to be recorded. One PM program evaluation revealed that the oil level of a large transformer was very low. When the PM sheets were inspected, they indicated that the level was normal. During an interview with the technician who performed the inspection, it was stated that a very low oil level was normal for that transformer. Facility management changed the inspection form to reflect the gauge markings (that is, Hi, Mid, and Low).

Don't assume a corrective action work order was written by the technician. Provide a space on the inspection form to record the work order number.

PM Development Teams

The importance of PM is usually understood by all in a facility, but operations people may balk when it comes to shutting off the equipment to do the work. It is helpful to involve all affected departments in identifying PM work, as well as developing the procedures. A *PM development team* is one way to involve all parties.

PM development teams are usually made up of one person from *operations,* one or more *technicians,* and one *maintenance management* person. Membership on the team should be rotated to other employees in these areas.

The agenda for the first meeting should include a general discussion of what constitutes PM work and an explanation of how to write a PM procedure. Each member should be provided with PM worksheets and given an assignment to write at least one procedure by the next meeting. All PM procedures should be reviewed, modified, and adopted in subsequent meetings.

Computerized PM Systems

A key to ensuring PM work is performed on time is to establish a method by which work orders are generated or action is triggered when the frequency has elapsed. Many manual methods have been used for this purpose, including *elaborate calendars* and *card filing systems*.

With the recent surge in the application of personal computers, many maintenance departments have implemented computerized PM systems, or a PM module in a CMMS. These systems automatically generate work orders based on the last date the PM work was performed plus the frequency. These systems can also provide detailed PM procedures. Some systems provide the option of predefined PM procedures.

Understanding Predictive Maintenance (PDM)

A successful PDM program relies on a dedicated effort to detect, analyze, and correct problems before failure occurs. As shown in Figure 4-5, the process consists of the following:

* *Periodic monitoring*—Once a new piece of critical equipment has been added to the program and baselined, it enters the PDM cycle. Measurement of the established parameters is taken periodically (weekly, biweekly, monthly, and so on).

* *Analyzing measurements that exceed the engineering limit*—If the measurement exceeds an established engineering limit, it must be analyzed further.

* *Problem analysis*—Analysis can take many forms. For example, a vibration signature can be taken on rotating equipment. A trained analyst may review for common problems (such as misalignment and imbalance), as well as the not-so-common problems (such as resonance).

* *Corrective work order*—Once the root source of the problem is determined, the best repair activity can be chosen. If the engineering limit is set low enough, there will still be plenty of

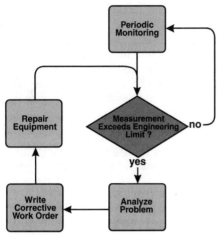

Figure 4-5   The PDM cycle.

time to correct the problem before further damage occurs. A work request is usually written to start the repair process.

- *Equipment repair*—Correction of the root problem will allow the equipment to re-enter the periodic monitoring program.

Spectrum of Predictive Maintenance

As stated previously, vibration measurement is probably the best known of current predictive applications. Table 4-2 shows the categories of industrial equipment that also benefit from a predictive approach.

Each failure mode can be linked to specific causes. The causes, in turn, can be detected by measuring certain physical parameters.

There has been a historical misconception that equipment failures cannot be predicted. Unexpected failures are largely attributed to Murphy's Law. With predictive technology, a vast amount of equipment failures *can* be predicted. In many cases, the *root problems,* which will ultimately cause the failure, are identified long before they cause a failure.

An effective way to better understand PDM is by examining a few examples. The following sections address vibration monitoring/analysis, lubrication analysis, and stress crack detection (acoustic emission).

Table 4-2 Spectrum of Predictive Maintenance

Equipment Category	Equipment Types	Failure Mode	Failure Cause	Detection Method
Rotating Machinery	Pumps, motors, compressors, blowers	Premature bearing loss	Excessive force	Vibration and lube analysis
		Lubrication failure	Over, under, or improper lube; heat and moisture	Spectrographic & ferrographic analysis
Electrical Equipment	Motors, cable, starters, transformers	Insulation failure	Heat, moisture	Time/resistance tests, I/R scans, oil analysis
		Corona discharge	Moisture, splice methods	Ultrasound
Heat Transfer Equipment	Exchangers, condensers	Fouling	Sediment/material buildup	Heat transfer calculation
Containment and Transfer Equipment	Tanks, piping, reactors	Corrosion	Chemical attack	Corrosion meters, thickness checks
		Stress cracks	Metal fatigue	Acoustic emission

Vibration PDM Programs

Research conducted during the late 1920s by Arvid Palmgren (an engineer with a bearing company) attempted to define the life of a bearing under various load conditions by experimentation and application of statistical methods. The result of tests on thousands of bearings provided an empirical formula for the life of antifriction rolling element bearings. For a ball bearing, the formula is as follows:

$$L^* = \left(\frac{C}{P}\right)^3 \times \frac{16{,}667 \text{ hours}}{\text{RPM}}$$

where the following is true:

- L is life of the bearing, expressed in hours.
- C is dynamic load rating of the bearing (lbf).
- P is actual load on bearing (lbf).
- RPM is speed of the rotating element (revs. per minute).

NOTE

This assumes proper lubrication.

The dynamic load rating (C) is determined by the bearing's geometry (physical dimensions) and the material used. The actual load (P) is the force the bearing sees in service. The asterisk (*) indicates that the formula only holds true if the bearing is properly lubricated. The life (L) is also referred to as the B_{10} life. The B_{10} life is the number of hours of operation before 10 percent of the bearings in a test group fail.

Palmgren's equation for bearing life is a standard today. A rotating machinery designer will use this formula to determine the life of bearings in a machine. An acceptable life for the machinery is chosen by the designer, based on production capability and market requirements. An estimate of the actual load (P) is made from installation assumptions and field limitations. Transposing the equation, the designer finds an acceptable dynamic load rating (C). Bearing manufacturers' catalogs contain the C value for all bearings they manufacture. Choosing the right bearing becomes a simple matter of matching the calculated value with those published in the catalog.

One important aspect of the bearing equation is the cube operator. If the *forces on the bearing are doubled,* the bearing life is not cut in half, but actually *reduces the life by a factor of 8* ($2^3 = 2 \times 2 \times 2 = 8$). For example, take a bearing in a machine that is sized for 10 years of life (87,660 hours). The life is reduced by a factor of 8 when the forces are double the design. The life expectancy of this bearing has now been reduced to a little more than a year (10 years divided by 8 equals 1.25 years).

Experiments in the 1930s showed that measurement of forces on bearings could be essentially accomplished by measuring the total movement of the machine during operation and measuring the speed of this movement. This movement is called *vibration*. Force levels on bearings can be determined by measuring the *vibration velocity* in inches per second (ips) or millimeters per second at or near the bearing points. Vibration velocity is also a good measurement of the *severity* of a particular vibration problem for equipment operating between 600 and 3600 rpm. *Displacement,* measured in mils, is a common measurement on slow-speed equipment operating at 600 rpm and below. It is also a good idea to take displacement readings on equipment with plain (non-rolling element) bearings. Proximity probes are commonly mounted close to the shaft on turbine generators and large fans for this purpose. This method is used rather than measurement at a bearing housing because much of the higher frequency vibration is dampened by the oil film in plain bearings.

Acceleration readings, measured in *gravities* (g), are more sensitive to higher-frequency vibrations. Whereas velocity is the change in displacement with respect to time, *acceleration* is the change in velocity with respect to time. For this reason, acceleration is often measured on high-speed equipment and for analysis of components that give off high frequencies (such as gears and bearings). Acceleration measurements are commonly used on equipment operating above 3600 rpm.

There are other vibration measurements used by different manufacturers to indicate higher frequency vibration or bearing problems directly. Shock pulse, spike energy, high frequency detection, and bearing fault detection are common proprietary readings different manufacturers use.

Defining Vibration Velocity Limits
Returning to our vibration example, let's review how a limit for mechanical rotating vibration measured on bearings was determined.

Vibration industry groups have conducted a number of experiments comparing the general condition of bearings to the vibration velocity levels on that equipment. These studies resulted in a standard *destruction curve* for bearings in certain classes of equipment. If increasing vibration velocity levels are left uncorrected, the result shown in Figure 4-6 can be expected for certain classes of equipment.

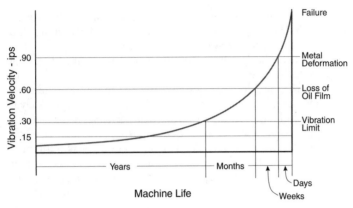

Figure 4-6 The bearing destruction curve.

The vibration velocity level on new equipment usually starts below 0.15 ips. The bearings should have a number of years of good life at or below this level. As the vibration reaches 0.3 ips, the equipment should be scheduled down to limit or eliminate the source of the vibration. If nothing is done, the bearings will fail in a matter of months.

If the vibration levels are allowed to increase until they reach 0.6 ips, the forces on the bearings are so severe that the oil film between the bearing balls and races breaks down and becomes ineffective. As stated previously, Palmgren's equation ceases to hold true if a bearing is not lubricated properly. As a result, bearing component wear is greatly accelerated.

If no corrective action is taken and the vibration level reaches 0.9 ips, the bearing vibration becomes audible. The force of the component contact is now high enough to exceed the elastic limit of the bearing metal, as defined by Hooke's law. At this point, metal deformation occurs and pieces of the bearing begin to break and flake off. The bearing's remaining life is now measured in days.

The Vibration Institute (a not-for-profit professional organization) and other organizations have established levels of equipment health as a function of vibration velocity based on experiments such as these. A simplification of this equipment health data is shown in Table 4-3.

Each piece of equipment in a monitoring program can be placed into one of these categories. This table is very useful when

Table 4-3 Rotating Machinery Ratings

Rating	Vibration Level	Necessary Action
Good	Less than 0.15 ips	Continue to trend
Fair	0.15 ips to 0.30 ips	Continue to trend
Poor	0.30 ips and above	Analyze and correct

categorizing vibration levels on most industrial equipment operating between 600 rpm and 3600 rpm.

Lubrication Analysis

Analysis of lubrication oils is a technology that emerged in the late 1940s. Originally pioneered by the North American railroads as a means of detecting bearing failure, it soon was applied by the United States Department of Defense for monitoring Air Force and Naval equipment.

The most common method of lubrication analysis uses atomic emission spectroscopy or *spectrographic analysis*. The presence and quantity of each element in the oil can be identified using an inductively coupled plasma (ICP) spectrophotometer. Another method growing in use is ferrographic analysis, which is a microscopic inspection of the wear particles in the oil.

A complete analysis of lubricating oil would include the following tests:

- Detection and measurement of *wear metals* (spectrographic)
- Detection and measurement of *additives* (spectrographic)
- *Physical tests* (spectrographic) to determine the lubricant serviceability
- *Ferrographic analysis* to examine extent and type of wear

In lubrication analysis, the presence of wear metals is of primary interest. If the amount of such metals becomes abnormal, it is an indication that excessive wear is taking place. The following list shows common wear metals along with common uses:

- Aluminum
  - Bearing caps
  - Thrust washers and pump bushings
  - Pistons and some engine blocks
  - Aluminum housings, impellers, or rotors

- Chromium
  - Shafts
  - Gears
  - Tapered roller bearings
  - Exhaust valves and cylinders
- Copper
  - Plain bearings (bushings)
  - Bearing retaining rings
  - Thrust washers
  - Oleate oil additives
- Iron
  - Gears
  - Bearings
  - Engine blocks, camshafts, rods, cylinders, crankshafts, valves, piston rings
- Lead
  - Babbitted bearings
  - Tetroethyl fuel additive
- Nickel
  - Steel alloys
  - Gear plating
  - Shafts
  - Rolling element bearings
- Silicon
  - Dirt
  - Silicon sealant
  - Antifoam additive
  - Antifreeze additive
- Silver
  - Rolling element bearings
  - Solder
- Tin
  - Rolling element bearings
  - Part of bronze alloys

- Zinc
  - Antiwear additive

Most lubricants have additives that enhance the service of the oil. These additives can retard oxidation, improve cohesion, or increase lubricity. The presence of additive elements can also confirm the type of lubricant specified in the sample data sheet. If the additive elements and quantities don't compare to the additive elements that should be in the oil specified, contamination of the oil is suspected. The following are common additive elements, along with their common uses:

- Boron
  - Dispersant additive
- Calcium
  - Detergent additive
- Magnesium
  - Machinery housings additive
- Molybdenum
  - Grease additive
  - Steel strengthener
- Phosphorus
  - Antiwear additive
- Potassium
  - Antifreeze additive
- Sodium
  - Dirt
  - Salt
  - Antifreeze additive (should not be in all)
  - Lubricant
  - Alkalinity improver

The physical tests can detect contamination or dilution with other oils. These tests basically monitor the serviceability of the oil. The tests performed include the following:

- *Water percent*—Moisture in oil
- *Carbon buildup*—Indicating oil breakdown
- *Viscosity*—Oil too thin or too thick

- *Silica*—Dirt contamination
- *Total acid number* (TAN)—Indicating acids in oil

Ferrographic analysis inspects the physical size and shape, as well as the size distribution of wear metals in lubricating oil. This inspection is most useful in the early life of the machine as defects (which will likely cause an infant mortality) can often be detected before a failure occurs.

The value in oil analysis comes most from regular monitoring. Although a single analysis may identify machinery problems, those problems are best detected as a trend identified from several analyses. The growing presence of iron, for example, would raise the suspicion of gear wear. Growing levels of tin and lead in a machine with babbitted bearings would point to bearing wear.

Stress Crack Detection (Acoustic Emission)

Another predictive technology allows the detection of early stress cracks in vessels, reactors, storage tanks, and pipeline transmission systems. The detection method is called *acoustic emission.*

As any containment structure goes through normal stressing (temperature changes, filling, emptying, pumping), any discontinuities in the structure material give out minute stress signals. The signals are relatively high in frequency, and although low in intensity, do travel quite well through the structure material itself. In acoustic emission detection, these signals are measured by highly sensitive tri-axial (detect in three planes) transducers. These signal inputs are fed into a computer. The computer, in turn, is programmed with a mathematical representation of the structure, along with the physical location of all transducers.

Through triangulation, the computer is able to detect the point of origin of the stress crack-induced signals. These can be further inspected through radiography to determine if the cracks have grown to the point of causing leaks or a rupture.

Although the technology of acoustic emission has not matured to the level of setting engineering limits of such signals, the method is being used successfully in monitoring critical equipment. (It is being used regularly on the Trans-Alaskan pipeline.)

PDM Success Stories and Failures

The results of vibration levels taken on thousands of pieces of machinery are shown in Table 4-4. The evaluation average is a good representation of the general condition of equipment across a broad spectrum of industry in the United States today. It represents data

Table 4-4 Rotating Machinery Condition

	Evaluation		
Rating	Average	Worst	Best
Good	42%	10%	43%
Fair	32%	25%	48%
Poor	26%	65%	9%

taken from process plants, steel mills, utilities, manufacturing plants, hospitals, and other institutions.

The evaluation indicates that more than one-fourth of the machinery in use today is in rough or poor condition with only a third of it in fair condition. The worst and best columns represent two distinct process plants. In 1981, both sites initiated PDM improvement programs. At that time, both plants were spending approximately $2 million annually in maintenance and repairs. Their initial machinery condition ratings were also close to the evaluation averages. The best plant backed the program solidly, and 5 years later had expanded the scope of the effort to include virtually all rotating equipment at the site. In 1986, their maintenance costs were only 72 percent of their 1981 levels. The worst plant, on the other hand, did not back the effort and dropped the program after 3 years. They felt that the time spent in monitoring could be better utilized fighting day-to-day problems. Their maintenance costs in 1986 were double the 1981 levels, and rising steadily.

Vibration levels indicate machine condition through the overall vibration velocity levels. Analysis of the vibration signal itself will help identify the root cause(s) of the vibration.

Properly applied, the analysis tools for rotating machinery can invariably detect machine problems long before failure. In the rotating machinery condition evaluation shown previously, the problems shown in Table 4-5 were detected in all machines that were determined to be in poor condition.

Table 4-5 Root Cause Machinery Problems

Cause	Evaluation Average
Misalignment	42%
Imbalance	33%
Damaged Bearings	32%
Internal/External Looseness	14%

Misalignment (of driver to driven machine) is consistently the most common machine problem detected. It is also one of the easiest problems to correct. It usually only requires some basic training for maintenance personnel in good alignment methods. Although damaged bearings are the third most common machine problem, they are usually caused by one of the other problems on the list.

Correcting Common Problems

Often, when a plant or facility is in the early stages of a vibration program, the mass of information collected by the vibration data logger can be overwhelming. A good first step is to determine which pieces of equipment are problems and which are not. The vibration limit of 0.3 ips is a good break point to use. All equipment 0.3 ips or higher should be repaired.

If the number of pieces of equipment at or above 0.3 ips is too large to tackle, a further prioritizing process may be necessary. Breaking down the equipment over the limit by type may be a good next step. Figure 4-7 shows a breakout by equipment for one facility.

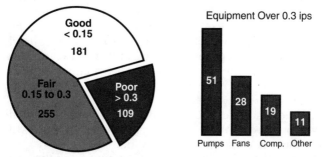

Figure 4-7   Categorized equipment failures.

Of the 109 machines over the limit, pumps and fans seem to have the biggest problems. Looking deeper into the work performed on pumps and fans can often reveal a facility-wide or systemic problem. For example, the facility that reported this data further investigated pump problems. The result indicated that there was little or no alignment of coupled machinery being performed. Deeper investigation revealed that the plant's only dial indicator (used in precision alignment) was damaged and locked away in a machinist's toolbox. The facility purchased new indicators and trained every mechanic in proper alignment techniques.

As stated previously, the most common problems found in rotating machinery through vibration analysis are (in approximate order) misalignment, imbalance, bearing damage, and looseness.

Grouping all problems together can start to reveal other systemic problems. Investigations can help facility management focus on the top problems that can be solved, with additional training, new methods, or new equipment.

The Four Ts of Correction

The correction phase of a PDM program is one of the most important parts of the program. Too often, time is spent on downloading the mass of data to start the program, collecting vibration data on hundreds of machines, and printing out reams of paper listing the machines with problems, but not on actually correcting the problems. Those involved with the PDM implementation gladly take on the task of analysis, but may not be involved in the correction of the problems they have identified.

The problem with vibration correction at most plants boils down to the *employee/machine interface*. The employee requires four things, called the *four Ts*:

- Time
- Target
- Tools
- Training

Time

First, the employee must be given the *time* to perform the work. Alignment procedures, for example, can take from 2 to 12 hours to perform properly. If an operation does not shut the equipment down for the time required, the alignment would not be complete. If the job is not planned and coordinated with operations to allow the proper time for repair, it should not begin.

Target

A craftsman working on equipment must have a methodical approach to the job. The goal of any method should be derived from a standard with a *target* for the best-desired result. In the case of alignment, a target of 0.002 inch on the rim and face of a coupling is commonly used. The employee must be aware of the target and know how to get there.

Tools

To perform the job, the *tools* required to do it right must be available to the employee. Too often, employees are required to perform a job that requires precision measurement devices or other special tools that either are not available or are part of only a select few that are allowed to be used.

Training

Finally, the craftsman needs the *training* in the skills and methods required for common repairs derived from a PDM program. As stated previously, grouping the problems and then developing or purchasing spot training to solve those deficiencies is an important step to a successful program.

Summary

Preventive maintenance is basic maintenance performed on machinery or facilities at an established interval or frequency. The main purpose of PM is to extend the equipment life and ensure capacity. PM can also be applied to protect personnel or the environment. PM can be performed when the equipment is shut-down, or it can include adjustments made while the equipment is running. PM work can be anything from a simple meter reading to a major rebuild of equipment.

One improvement on PM is predictive maintenance (PDM). First considered an adjunct to most PM programs, PDM is now recognized as distinct from PM efforts. While PM tasks are initiated by a calendar (based on a set frequency), PDM programs measure and react to the condition of operating equipment. PDM compares the trend of measured physical parameters against known engineering limits for the purpose of detecting, analyzing, and correcting problems before failure occurs. Hopefully, the limit is low enough to detect the problem before excessive damage occurs. The repair process can be enhanced with a good analysis of the problem. The correction of the root problem is the key to most predictive efforts.

The first step in any maintenance effort is to meet mandatory PM requirements. Mandatory PMs are activities performed on equipment or facilities as required by law or contract. Discretionary PM activities performed on equipment or at a facility are based on the need to reduce downtime, minimize costly equipment damage, or ensure personnel safety.

Once a list of equipment requiring mandatory or discretionary PM has been defined, a solid PM procedure must be developed to combat the root cause of the problem. The types of PM procedures that should be considered include inspections, adjustments, testing, calibrations, rebuilds, and replacements. Sources that can help identify PM procedures include vendor-recommended PMs, plant experience, generic PMs, and equipment history.

A successful PDM program relies on a dedicated effort to detect, analyze, and correct problems before failure occurs. The process consists of periodic monitoring, analyzing measurements that exceed the engineering limit, problem analysis, corrective work order, and equipment repair.

There has been a historical misconception that equipment failures cannot be predicted. Unexpected failures are largely attributed to Murphy's Law. With predictive technology, a vast amount of equipment failures *can* be predicted. In many cases, the root problems, which will ultimately cause the failure, are identified long before they cause a failure. Common types of PDM programs include vibration analysis, lubrication analysis, and stress-crack detection.

The correction phase of a PDM program is one of the most important parts of the program. Too often, time is spent on downloading the mass of data to start the program, collecting vibration data on hundreds of machines, and printing out reams of paper listing the machines with problems, but not on actually correcting the problems. Those involved with the PDM implementation gladly take on the task of analysis, but may not be involved in the correction of the problems they have identified.

Chapter 5 examines a variety of common scheduling methods, including prioritizing maintenance work, using a maintenance backlog, the allocation scheduling method, CMMSs, and priority numbering systems.

Chapter 5

Scheduling Methods

Just as no one form of management is best for every facility, *no one scheduling method fits every facility*. Maintenance scheduling methods tend to vary depending on the type of operation. Institutions, hospitals, and commercial building maintenance departments are under different restrictions from process or manufacturing facilities. Central shop maintenance departments have different restrictions from area maintenance departments. Utilities with a multicraft workforce have more scheduling constraints than workforces with no craft distinctions.

Whenever any new system is implemented, the basic rules for the effort must be defined. These rules can be the groundwork for identifying the scheduling method that fits best in a facility.

This chapter examines a variety of scheduling processes and methods, including the following:

- Prioritizing maintenance work
- Using a maintenance backlog
- Allocation scheduling method
- Scheduling using a CMMSs
- Priority numbering systems

Prioritizing Maintenance Work

Bringing the most important jobs at a facility into the open is the key to ensuring the right work is performed at the optimum time. The process used to determine the relative importance of jobs should take into account the long-term and short-term requirements of a facility. This process is called *prioritizing*. The word prioritizing means different things to different people. One way to approach the true definition may be to ask the following question:

What is the highest priority job in your facility?

- *According to an originator* —The highest priority job is the highest priority job in the originator's particular area, regardless of priorities in other areas.

- *According to a maintenance supervisor*—The highest priority job is any emergency job that comes up that day, or the number one job on the schedule.

- *According to an operations manager*—The highest priority job is in a manufacturing area where an outage will cause a missed shipment or lost sale.
- *According to a plant manager*—The highest priority job is the one that has the highest environmental or safety risk.

It is also possible that the highest priority job in the plant may be none of the above. The individuals listed here have a valid reason to believe that their assessments of the most important jobs in the facility are correct. The supervisor and originator's perceptions of priority are more immediate. The priorities of the plant manager and production manager are more long-term.

The prioritization process begins as requests for maintenance work are received. It may be helpful to break down the perception of priorities into three basic stages:

- *First*—Priority with respect to classification
- *Second*—Priority with respect to requested completion date
- *Third*—Priority with respect to schedule

Classification

When a work request is first generated, it can be assigned a *classification*. As discussed in Chapter 1, the *class* of most maintenance work can be broken down as follows:

- Preventive maintenance (PM)
- Predictive maintenance (PDM)
- Prefabrication
- Corrective
- General maintenance

PM, PDM, and prefabrication work should initially be the highest priority work at the facility. Few facilities can afford to put off lubrication of equipment or other PM work, without paying for that decision in the long run.

Requested Completion Date

As the highest priority job, PM does not mean that all other work must wait until a PM job is completed. A custody plan dictates an optimal time this work can be performed to provide the maximum benefit to the facility. The *requested completion date* written on a work request helps indicate the urgency of the job with respect to all other jobs received by the maintenance department. The requested completion date is used by a planner as a goal for completing the plan and ensuring that all resources are available.

As the requested completion date of a job approaches, the relative importance of the job (when compared to other jobs in the plant) becomes clearer. When the job is eventually placed on a schedule, the priority can be redefined with respect to *schedule priorities*.

Schedule Priority
The importance of each job on the day it is scheduled can be broken down into two basic categories:

- *Planned work*—Any work where resource estimating and logistic restrictions dictate or allow the job to be placed on a schedule.
- *Emergency work*—Any job that displaces work from the schedule.

Planned Work
Of the planned work, PM, PDM, and prefabrication work have the highest priority, and should be started first. Corrective and general maintenance work have a lower priority.

Implied in the definitions for planned and emergency work is the fact that a schedule should be made up of only planned work and not emergencies. When an emergency occurs, one or more of the scheduled jobs must be delayed or stopped to complete the emergency. The first jobs to be dropped from this schedule are jobs with a general maintenance or corrective classification.

Emergencies
Emergency work is generally the most expensive kind of maintenance. Usually, the work is assigned without the benefit of a plan, and the parts required to do the job may not be available. The maintenance worker assigned the job must also do the planning on the fly. Numerous trips to the storeroom are also common when performing an emergency job, which adds to the downtime.

Figure 5-1 shows a hierarchy of the different types of emergencies that exist in many facilities.

The first category of emergencies is those that are *real*. This means something is broken and must be fixed. Either production or manufacturing output is compromised, or a grave safety or environmental condition exists. These real emergencies, in turn, consist of two types: those that are foreseeable and those that are not.

Unforeseeable emergencies are bona fide. There is no reasonable or economical way that the problem could have been detected

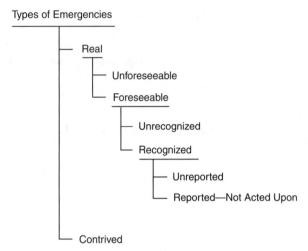

Figure 5-1 Emergencies explained.

before it became an emergency condition. *Foreseeable* emergencies, however, consist of two types: those that are unrecognized and those that are recognized. *Unrecognized* (but foreseeable) conditions (which could lead to an emergency) represent situations that require management direction or emphasis and sometimes training. Workers who do not recognize a potential emergency condition must be trained to see the little problems that will grow into big ones.

The *recognized* conditions also consist of two types: those that are reported and those that are not. *Unreported* (but recognized) emergencies represent a definite problem for management. Knowledgeable people do not feel compelled to report situations they recognize as potential problems. It is possible they feel that nothing will be done if they are reported. This points to the last category of real emergencies—those that are *reported* but not acted upon. This category is the worst kind of emergency. They are the fault of the maintenance department. They undermine the credibility of the maintenance effort and, in fact, create the other type of emergency: condition emergencies that are *contrived*.

Contrived emergencies aren't really emergencies. Instead, they are an abuse of the priority system. They come about when the originator believes that the job will not be started unless it has a high priority.

Emergencies are expensive and often outside the control of the maintenance effort. But one action that maintenance has total control over is to see that no reported pre-emergency conditions go uncorrected. Building credibility takes time, but it is the only way that contrived emergencies will be eliminated.

As the maintenance effort ensures that *no* reported pre-emergency conditions go uncorrected, more foreseeable emergent conditions will be reported. As these, in turn, are corrected, better surveillance of operations will become the norm, and emergencies will only be *real*.

Using a Maintenance Backlog

The list of work generated from maintenance requests is called the *backlog*. This list will grow or shrink as jobs are added or work is completed. Other work (such as emergencies or work that has not been identified with a written work request) is not part of the backlog.

As the days progress, a work order in the *backlog* may not be finished or scheduled in time and, hence, *age beyond the requested completion date*. The job may no longer be valid, and the originator should remove it from the backlog. On the other hand, the job may still be valid but the priority with respect to the requested completion date is no longer valid. To ensure that the backlog of work is valid and reflects the actual needs of the facility, steps must be taken to control the backlog.

Controlling Backlog

The backlog of hours can be an important tool to determine what *resources* are required in the coming weeks, months, and years. An index that provides a way to look at the backlog, expressed in weeks, is expressed as follows:

$$\text{Backlog} = \frac{\text{Hours of Schedulable Work}}{\text{Hours in a Work Week} \times \text{Numbers of Available Workers}}$$

The calculation of backlog weeks provides one guide for the maintenance manager to help make adjustments in the maintenance workforce. A common guide used by many maintenance managers requires that the backlog fall within the range of *2 to 8* weeks.

If the backlog falls below two weeks, efficient scheduling of the workforce is difficult. The most common cause of low backlog is the failure to identify valid work in advance. This usually results in a high emergency rate.

If the backlog is consistently below two weeks, but the emergency rate is low, the workforce may be too big. In this case, cutbacks in the number of employees performing maintenance work may be a necessity.

If the backlog exceeds 8 weeks, high-priority work may not be getting done in time. Temporary increases in overtime or the use of contract labor may cure this problem in the short-run. If the backlog consistently exceeds 8 weeks, a permanent increase in the workforce may be necessary. Productivity improvements (such as new tools or improved planning) are also good ways to reduce the backlog.

Invalid Backlogs

A backlog of maintenance work should reflect all the valid work the maintenance department must complete. Too often, the backlog of work requests includes jobs that are no longer valid. In other words, the work requests no longer represent work that will be required to keep the facility running or maintaining capacity. This usually occurs as facility priorities change. If one intention of the backlog calculation is to make decisions about the size of the workforce, the validity of that calculation must be ensured.

To ensure the backlog remains valid, a dialog must begin between maintenance, operations, and engineering. One point at which this dialog starts is when the requested completion date suggested by the originator has expired. When this occurs, it is the responsibility of the maintenance department to inform the originator and request a new date. This communication helps determine the priorities of most maintenance work and helps ensure the validity of the backlog.

One good way to control the backlog is the use of *scheduling meetings*. These meetings are the forums by which originators or their representatives decide what work requests will be performed in the following week or month. Also discussed is the viability of work orders.

Calendars

Prior to using any scheduling system, a list of nonwork events that will affect that schedule must be compiled. These events include *holidays, vacations,* and *meetings* that occur during the year. These events should be laid out on a *calendar* manually, or entered into a CMMS calendar. Writing down the total available workforce for each day or week in the year on the calendar helps pick out low workforce periods so that required changes can be made.

An *operations schedule* should be provided to maintenance indicating the times when maintenance can have custody of an operation area or equipment in the coming week, month, and year.

A *weekly calendar* must then be built prior to schedule development. The resulting calendar will be the workforce list used when scheduling work.

Scheduling with a Plotted Backlog

One tool that can be very helpful to the scheduler when developing a daily or weekly schedule is the plot of available work hours. The steps to doing this begin with the following:

1. Put all planned work orders in order by required completion date.
2. Group the work orders by the week they must begin to meet the required completion date.
3. Total all hours each week and plot the data.

This procedure should be repeated for each craft in multicraft organizations. Figure 5-2 shows an example of a typical plot. This plot shows the hours needed to be completed by the maintenance department. A small amount of overdue work is acceptable, but excessive

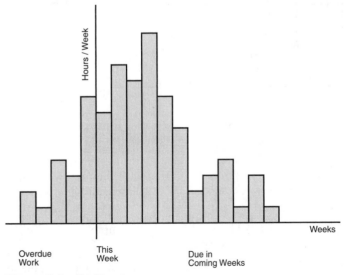

Figure 5-2   Backlog 1.

overdue work must be reprioritized. The planner may need to renegotiate a new required completion date with the originator. Some
planners have the authority to change the priority on their own.

It may be a good idea to remove any PM work from the backlog
before making this plot. Often, PM work orders are generated by
computerized systems the week before they are due. Hence, the actual PM workload for all the weeks displayed is not reflected in the
plot. The PM work will be taken into account later in this procedure.

An option to backing out the PM work is to generate all the PM
work for all the weeks displayed on the plot. This is not always
possible, so most people opt to back it out.

After reprioritizing the overdue work and backing out PM, a new
and improved plot should be made using the same rules. The cleaned
backlog may look like the plot shown in Figure 5-3.

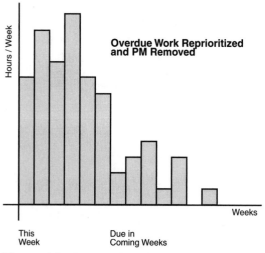

Figure 5-3   Backlog 2.

Next, the number of hours associated with the total workforce
working a normal week should be calculated and plotted. For
example:

85 employees × 40 hours per week per employee
= 3400 hours per week

This means that there are only 3400 hours of labor available in
any week without overtime. It should be noted that this calculation

can be replaced by plotting the weekly labor calendar instead. However, simple calculation will suffice for this example.

The total workforce should be adjusted for absenteeism. The result is represented by the following formula:

$$3400 \text{ hours} \times \frac{100\% \text{ attendance} - 7\% \text{ absenteeism}}{100} = 3162 \text{ hours}$$

This number must be adjusted for the workforce that usually works on PM jobs, since they are no longer in the plot. Referring back to the previous example, the average PM work performed in a week is 822 hours, so the calculation would be as follows:

$$3162 \text{ hours} - 822 \text{ hours of PM} = 2340 \text{ hours}$$

These numbers are plotted in Figure 5-4.

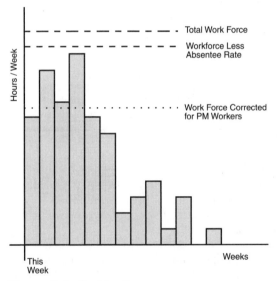

Figure 5-4 Backlog 3.

A review of this chart reveals some important characteristics. The workload in coming weeks exceeds the available workforce. Even with overtime, the maintenance department would not be able to meet the priority requirements of the next few weeks.

This problem is compounded by the fact that emergency work and other high-priority work will arise in the coming days and weeks. If the workforce is not capable of handling the normal work load, even with overtime, there will be no way for much of this work to be completed in the coming weeks.

The solution is reprioritization of all jobs in the workload. The resulting plot is shown in Figure 5-5.

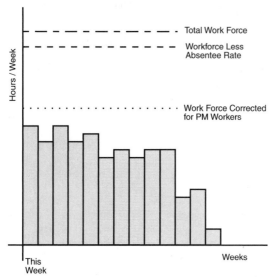

Figure 5-5 Backlog 4.

With the backlog reprioritized, the planner/scheduler can feel confident that a reasonable schedule will be developed each week. Building a schedule with work that has been reprioritized is very easy. There may even be opportunities to do some work long before its required completion date.

The backlog plot should be developed each week by the planner to identify potential staffing problems and the need for reprioritization. Many CMMSs have this capability in some form. Use of PC-based spreadsheet programs can greatly enhance the planner/scheduler's ability to develop backlog plots. CMMSs that do not have graphic capability often have the capability to download data into many spreadsheet formats for further analysis.

Allocation Scheduling Method

Facilities are often split into processing segments or divided by types of operation (such as utilities, manufacturing, or packaging). Responsibility for each area may be assigned to an operations supervisor or team leader. Maintenance in these segments is often provided via an *area maintenance* approach. A separate (and often autonomous) maintenance workforce is assigned to each operating area.

One argument for the area approach is that the maintenance workforce can be staffed and specially equipped to deal with the immediate needs of an operating area more quickly. Some maintenance managers complain that the emphasis on immediate needs often means that PM and PDM work receives a lower priority. Additionally, some areas of the facility are starved of workers while others have more than they need.

The *allocation scheduling method* is most effective in facilities that use the area maintenance concept. This approach refocuses the workforce on PM and PDM work, while providing a facility-wide priority system to complete the other important jobs in the facility. The method lets each operations supervisor or team leader determine the corrective or general maintenance that should be scheduled on a weekly basis. The allocation method is also successful when more than one operations area within the facility is vying for the same maintenance workforce.

The following describes the steps to implement a typical allocation scheduling system:

1. The planner/scheduler builds plans for jobs in the backlog, ensuring that parts and equipment are available to start the job on or before the requested completion date.

2. On Thursday of every week the total workforce for the coming week is determined, based on vacation and holiday schedules, as well as the average absenteeism for the facility. The result from a typical facility is shown in Table 5-1.

   Labor hours required for PM and PDM work for the next week are totaled and subtracted from the total labor hours. This number is recorded as the total *allocation* of labor hours for each area.

3. An initial estimate of workforce needed in each area is calculated based on planned labor hours completed in the previous week, as shown in Table 5-2.

   Adjustments to allocations are made during a negotiation process between the operating areas at a Friday scheduling meeting (to be described later).

4. A backlog of planned work is compiled for each area. It is then distributed to area operations supervisors, superintendents, and managers, along with the total allocation for each area. A sample backlog for Area D is shown in Table 5-3.

Table 5-1 Sample Facility——Available Workforce: Week of 7/7/1997

Item	Detail	Entry	Total
Total Workforce		55	
Vacations Scheduled		−5	
Net Available Workforce		50	
Total Available Labor hours	50 × 40 hrs/week		2000
PM/PDM Scheduled			
	Area A	133	
	Area B	138	
	Area C	66	
	Area D	52	
	Area E	111	
Total PM		500	−500
Net Available Labor Hours			1500

The backlog lists the total labor hours remaining to complete each job and the crew size for each job. The value of labor hours for each craft should be broken out in a multicraft facility. The operations supervisors in each area should review the backlog and check off jobs that should begin next week. In the case of a multicraft facility, each craft segment of a job or corresponding job must be checked off as well.

The labor hours for these jobs should be totaled and compared to the allocation for the area. The day or days the job should be worked are also noted. Some large jobs may run over into the subsequent weeks. In this case, the planner and operations supervisor must note how many labor hours should be worked next week.

Table 5-2 Sample Facility——Initial Allocation

	Allocation % (from history)		Schedulable Hours
Area A	33%	× 1500 =	495
Area B	23%	× 1500 =	345
Area C	23%	× 1500 =	345
Area D	12%	× 1500 =	180
Area E	9%	× 1500 =	135
Total	100%		1500

Table 5-3 Sample Facility——Ready Backlog for Area D as of 7/4/97

W/O#	Date Initiated	Date Req.	Equip #	Est Hrs	Description	Selection
02908	12/26/96	3/12/97	05P102A	6	Install new drip pan.	
03172	3/10/97	5/9/97	05T905	40	Fab new manhole cover for T-905.	
03350	5/26/97	6/2/97	14BLDG	4	Paint "QUALITY" sign.	
03426	6/23/97	6/26/97	47R200	14	Repair minor leak at reactor seal.	×
03357	5/28/97	6/27/97	37R907	8	Remove knock-out pot.	×
03398	6/13/97	6/28/97	37R907	7	Replace rupture disc with lower pressure.	× 7/11
03402	6/15/97	6/28/97	05P201	2	Repair stuck drawer.	×
03364	5/31/97	6/29/97	05BLDG	16	Annual shop floor treatment.	×
03369	6/2/97	7/1/97	05T905	6	Replace frayed hose.	×
03414	6/19/97	7/4/97	26B501	11	Increase fan speed, change sheave & motor.	× 7/9
03415	6/19/97	7/4/97	10B501	9	Replace fan bearings—per vibration analysis.	× 7/8
03450	6/30/97	7/7/97	37C031	4	Balance compressor lobes—per vib. analysis.	× 7/11
03430	6/24/97	7/8/97	10M210	11	Replace worn mixer blades.	× 7/8
03339	5/21/97	7/10/97	14PK205	15	Fabricate and install new bagger guide.	×

(continued)

143

Table 5-3 (continued)

W/O#	Date Initiated	Date Req.	Equip #	Est Hrs	Description	Selection
03399	6/13/97	7/12/97	37BLDG	8	Repair damaged insulation.	
03447	6/29/97	7/14/97	26G204	5	Change oil in gearbox—per lube analysis.	× 7/9
03458	7/3/97	7/17/97	37P303B	6	Cut down impeller size.	×
03351	5/26/97	7/24/97	05V108	14	Paint vessel and piping.	×
03372	6/2/97	8/1/97	26T200	6	Replace T-200 modify pipe.	× 7/9
03454	7/1/97	8/13/97	37P201	25	Replace guard and nipple.	×
03353	5/27/97	8/24/97	26T607	6	Weld tank crack.	× 7/9
03459	7/3/97	9/7/97	37P031A	24	Install hot oil disc.	
03456	7/2/97	9/23/97	10I210	5	Install & calibrate photometric analyzer.	
03465	10/13/97	10/13/97	37T973	15	Relocate level controls and dipleg.	×
03462	7/4/97	11/2/97	10K204	16	Install steam tracing.	
03424	6/22/97	11/25/97	05M121	2	Check and/or repair lubricator.	
03440	6/27/97	12/20/97	18C220	25	Install new lights.	

Table 5-4 Sample Facility—PM and PDM Work Due the Week of 7/7/97

W/O#	Equip #	Est Hrs	Description
P14315	26C420	4	PM—Compressor quarterly.
P14318	GEN	11	PM—Greasing route D4.
P14321	GEN	16	PM—Oil route D5.
P14322	MOB23	2	PM—Pickup truck safety check.
P14324	GEN	7	PDM—Vibration monitoring—Week 27.
P14326	37T910	4	PM—E13D transmitter calibration.
P14327	47R100	8	PM—Perform leak test on reactor.
	Total	52	

5. A list of the PM and PDM work due to be performed should also be provided, as shown in Table 5-4.

6. A meeting of operations representative(s) and maintenance planner/scheduler(s) is held on Friday. The operations personnel provide a marked up backlog for their area. The planner/scheduler verifies that the allocation for each area has not been exceeded. During the meeting, the planner/scheduler should also bring to light any jobs with a requested completion date that may expire if they are not started next week. A request should be made for these jobs to be redated, rescheduled, or removed from the backlog.

7. Sometimes work required by an area exceeds the allocation for that area. In these cases the area supervisor can request another area supervisor give up one or more jobs and allot the labor hours to that area. If they cannot reach a decision, they must involve the operations manager or facility manager as the mediator. The planner/scheduler will recalculate the workforce allocation based on operations management decisions (see Table 5-5).

8. A finalized list of all the work chosen by operations as well as PM and PDM jobs to be worked next week is provided to all operations managers by Friday afternoon. The planner/scheduler takes the total labor hours for each area and determines the optimum staffing required for each day of the next week.

Table 5-5 Sample Facility—Final Area Allocation Adjustments

		Initially Chosen (by Area)	Final Required (by Area)
Area A	495	470	470
Area B	345	345	310
Area C	345	405	405
Area D	180	180	180
Area E	135	135	135
Total	1500	1535	1500

Weekly Schedules

The result of any scheduling system is a *weekly schedule*. An example of a typical weekly schedule is shown in Table 5-6. In this example, the following assumption is made:

Area D: Available Hours:
180 hours + 52 PM hours = 232 hours

The schedule was developed on the Friday before (7/4/97) with the intention of providing a week's worth of work for each employee in Area B.

Scheduling for a full workweek without accounting for breaks in the schedule is a good basis for developing a weekly schedule. Even though it is inevitable that at least one job will be put off for an emergency, the job that was put off can be considered to be the lowest-priority job on the weekly schedule. It is important to note which jobs are dropped from a schedule in lieu of a higher-priority job that may come up during the week. Maintenance personnel may choose to work the displaced job on overtime, or may even assign a contractor to do the job if it is still deemed important. Operations may decide that the job was not as important as they originally thought it was, and put it off until next week. This, in essence, reprioritizes the job.

In the example, the estimated hours for some of the jobs exceed the amount of work that can be economically completed in a normal workweek. As a result, the planner/scheduler and operations representative have determined a reasonable amount of labor hours that can be completed during the week.

Some equipment in the plant may be scheduled down by operations. To take advantage of this downtime, some jobs were marked with the date for this scheduled downtime.

Table 5-6 Sample Facility—Weekly Schedule Beginning 7/7/97

W/O #	Equip #	Description	Est Hrs	S/D Date
P14315	26C420	PM—Compressor quarterly.	4	7/9
P14318	GEN	PM—Greasing route D4.	11	
P14321	GEN	PM—Oil route D5.	16	
P14322	MOB23	PM—Pickup truck safety check.	2	
P14324	GEN	PDM—Vibration monitoring—Week 27.	7	
P14326	37T910	PM—E13D transmitter calibration.	4	7/11
P14327	47R100	PM—Perform leak test on reactor.	8	
03426	47R200	Repair minor leak at reactor seal.	14	
03357	37R907	Remove knock-out pot.	8	
03398	37R907	Replace rupture disc with lower pressure.	7	7/11
03402	05P201	Repair stuck drawer.	2	
03364	05BLDG	Annual shop floor treatment.	16	
03369	05T905	Replace frayed hose.	6	
03414	26B501	Increase fan speed, change sheave & motor.	11	7/9
03415	10B501	Replace fan bearings—per vibration analysis.	9	7/8
03450	37C031	Balance compressor lobes—per vib. analysis.	4	7/11
03430	10M210	Replace worn mixer blades.	11	7/8
03339	14PK205	Fabricate and install new bagger guide.	15	
03447	26G204	Change oil in gearbox—per lube analysis.	5	7/9
03458	37P303B	Cut down impeller size.	6	
03351	05V108	Paint vessel and piping.	14	
03372	26T200	Replace T-200 modify pipe.	6	7/9
03454	37P201	Replace guard and nipple.	25	
03353	26T607	Weld tank crack.	6	7/9
03465	37T973	Relocate level controls and dipleg.	15	
		Total	232	

Operations should get a copy of the weekly schedule to help arrange equipment shutdowns and other coordinations with the maintenance workforce.

Daily Schedules

Once the schedule is developed, some planners/schedulers simply turn over the weekly schedule to the maintenance supervisor and wash their hands of what happens from day to day. In these cases, the maintenance supervisor must determine which jobs should be worked each day, and must also coordinate equipment shutdowns with operations. The supervisor must also make decisions about new and pressing jobs that may come in during the week. This can add a burden to the supervisor's job that is really unnecessary.

Many planners/schedulers develop a schedule for each day to free the supervisor from some of the coordination requirements. The form shown in Figure 5-6 is typical of a daily schedule used at many facilities.

The schedule is developed for the next workday (Tuesday) and for a specific area (Area D). Operations should review the schedule to determine which equipment must be shut down for maintenance. The first two jobs on the schedule require a shutdown discussed with operations the week before. Permitting and lock-out requirements can also be discussed at this time.

In the example, an effort was made by the planner/scheduler to provide a day's work for each maintenance employee in the area. The normal staff was listed along the top of the schedule. An employee available from another area was added (by hand) to help finish the schedule on this day.

Some information has been added to the schedule to help anyone who might want to audit it later. Using the CMMS, the planner/scheduler has listed the number of times each job has been placed on a schedule. The original estimate (Orig. Est. Hours) and the hours already charged to the job (Compl. Hours) are also listed.

It is usually the responsibility of the team leader or maintenance supervisor to assign the right person or persons to each job. The team leader or maintenance supervisor also has the responsibility of making every effort to start each job on the schedule.

In the example, the marked-up schedule indicates the results of the day's effort. Equipment was shut down for the first two jobs on the schedule, so these jobs were started first thing and worked until completed. When these shutdown jobs were finished, the maintenance employees were moved to complete a lubrication PM and a continuation of a job to relocate some level controls.

Daily Schedule

Area	Day	Date
D	Tuesday	07/08/97

W/O #	Equip. #	Description	D. Huff	R. Smith	J. Stallcup	J. Bartlett	A. Meyers	A. Hanson	Days on Sched.	Orig. Est. Hours	Compl. Hours	Act. Hours Today
03415	10B501	Replace fan bearings		4.5			4.5		2	9	0	⑨
03430	10M210	Replace worn mixer blades	5.5		5.5				0	11	0	⑪
P14321	PLANT	PM - Oiling route D5	2.5		2.5				1	16	11	⑮
03465	37T973	Relocate level controls		3.5		3.5			1	15	8	⑦
P14318	PLANT	PM - Greasing route D4	X						1	11	0	0
P14322	MOB23	PM - Pickup truck safety check			X				0	2	0	0
03454	37P101	Replace guard and nipple			6				4	25	8	6
03481	35P201	EMERGENCY - Replace pump seal	10									⑩
03480	37M120	EMERGENCY - Replace sample valve			2							②
			Straight Time	8	8	8	8	8	8			NOTE: Circle Act. Hours for Completed Jobs	
			Overtime	2	0	0	0	0	0				

Figure 5-6 Daily schedule.

Two emergency work orders were covered and completed during the day. They were added to the schedule by hand and were performed in place of two scheduled jobs, which is understandable if they were true emergencies. One of the emergency jobs extended into overtime. When the two-hour emergency was completed, the maintenance employee was reassigned to a scheduled job.

The jobs that were not completed may be put on the schedule for the next day, or left for another day during the week.

Auditing a Completed Daily Schedule

A daily schedule can provide very good control over weekly schedule activity. Jobs on a weekly schedule are easily ignored in lieu of so-called emergencies. Maintenance supervisors are often persuaded to perform added jobs that were not discussed with the planner/scheduler. A completed daily schedule tells all. If a scheduled job has not been performed, a maintenance manager can ask some key questions.

The maintenance managers should ask questions such as:

> *Why wasn't this job started today?*

or,

> *Why wasn't this job completed today?*

The *answers* will often resemble these:

> *The parts and materials were not available.*
> *Operations didn't shut down the equipment.*
> *An emergency job came up.*
> *The estimate was too low for the work involved.*

These answers should evoke more questions, this time directed toward the responsible individuals. The maintenance manager may ask the planner/scheduler why parts were not available for a job on the schedule. The planner/scheduler should also be asked if arrangements were made with operations to have the necessary equipment shut down. The operations department should be questioned as to the validity of an emergency job, and why it was necessary to displace a scheduled job if it was not an emergency.

The maintenance manager may also ask about the progress of longer jobs that were not completed. Jobs that have been on the daily schedule a number of times and not completed should also be scrutinized. Some jobs may be put off by the maintenance supervisors because they wrongly feel they are not as important as others. This happens more often when the daily schedule includes many more jobs than can be completed within a normal workday.

If the actual time spent on a long job exceeds the estimate, the maintenance manager should determine the reason why. Both the planner/scheduler and the maintenance supervisor should be included in a discussion to determine if the problem is with the job performance or the estimate.

On the other hand, the maintenance manager should also ask questions if all the scheduled jobs were completed within the estimate. This is especially questionable if added jobs or emergencies were also completed during the day. The job estimates may be too high when this occurs.

The maintenance manager should correct this situation because questionable estimates call all other components of a scheduling effort into question. High backlogs that seem to indicate the need for a larger workforce are suspect. Operations may disbelieve the time estimates for downtime required on certain jobs. If this situation persists, weekly schedules and daily schedules will also be considered invalid by facility personnel.

Sample Maintenance Daily Scheduling Procedure

It's often a good idea to have a *written procedure* that describes the activities and responsibilities in a scheduling program. The procedures shown in Figure 5-7 can apply to the daily scheduling and execution of maintenance work.

Scheduling in Computerized Maintenance Management Systems (CMMS)

All CMMS programs that provide a scheduling module include *calendar* and *resource files*. The calendar file is nothing more than a schedule of workdays and, in some of the more sophisticated programs, it also provides information identifying the work hours in a workday. Holidays, other nonwork days, and the normal workday hours are identified for the year. The calendar is interactive with the scheduling module, in that changes within the calendar will immediately be reflected in the schedule. The ability to meet priority demands can be quickly viewed by adding Saturdays to the workweek, or changing the workday to a 10- or 12-hour length.

The resource file keeps track of workforce levels, critical support equipment, or other important constraints to any resources that must be tracked. Again, the resource file is interactive with the scheduling module. That is, as vacations cause fluctuations in

1.0 Execution of Work

1.1 The maintenance supervisor is responsible for the assignment of work from the daily schedule. The daily schedule shall indicate the employee(s) assigned to specific work orders.

1.2 High-priority work orders received at the beginning of a work day for immediate assignment shall be written onto the daily schedule.

1.3 Work orders that cannot be finished because parts are not available, equipment was not made available by operations, or other delays shall be returned to the planner/scheduler for reissue.

2.0 Daily Scheduling

2.1 Maintenance supervisors will meet their respective planner/scheduler at a mutually convenient time during the workday. All this meeting the supervisor will advise on the status of all work on the daily schedule.

This meeting will normally be held during the afternoon, allowing adequate time for the preparation, printing and distribution of the next day's or subsequent schedules.

The following information will be covered in this meeting.

2.1.1 *Completed Work*

The planner/scheduler should be informed of work orders that will be completed during the day. If there is doubt that the job will be finished it will be considered "Work In Progress".

2.1.2 *Work In Progress*

Work orders that have been assigned and are currently being worked are considered "Work In Progress". The anticipated hours that will be charged to each job during the workday will be conveyed to the planner at the meeting.

The planner/scheduler shall be advised of any high-priority work orders that are in progress and must be completed but will extend beyond the normal workday. The supervisor and planner/scheduler will discuss whether or not the job will be reassigned to a subsequent shift(maintenance swing or other temporary back shift assignment). The maintenance manager will be advised of any potential premium rate work that may occur.

2.1.3 *Work On Hold*

Work orders which have been assigned and started but cannot be completed will be considered "Work On Hold". Such work may be interrupted or delayed because parts are unavailable: equipment was not made available by operations, or any other delay. If such delay is anticipated to extend beyond 5 or more days, the work order status shall be changed and the job should be removed from the daily schedule. Such work will return to the planner/scheduler until it is ready for reissue.

2.1.4 *Head Count Adjustments*

The maintenance supervisor shall convey to the planner/scheduler any changes or adjustments to the crew staffing level. Sickness, vacations, absenteeism or any other reasons for changes to the crew size for the current day and subsequent days shall be used by the planner/scheduler. Personnel may be moved from one area to another or the planner/scheduler may choose to reduce the number of jobs on the schedule.

2.1.5 *Added Work*

The planner/scheduler and maintenance supervisor will discuss work orders to be added to the schedule for the next day. The maintenance supervisor will advise on "Work In Progress" that is nearing completion. The Planner will advise on backlog work orders that are a high priority and on "Work On Hold" that is available for reissue. Work orders to be added that require safety clearances will also be identified.

Figure 5-7 Sample maintenance daily scheduling procedure.

2.2	After the meeting, the planner/scheduler shall prepare a daily schedule for the next day (or shift). The planner/scheduler will select work orders from the current backlog, based on priority or date needed, to ensure at least a full day's work is scheduled for all maintenance employees.
	The work selected shall also reflect a mix of job sizes, allowing the maintenance supervisors flexibility during job assignment. Whenever possible, the additional work shall be selected in an attempt to minimize distance or travel time between jobs.
2.3	The planner/scheduler will publish and distribute daily schedule at least 1 hour before the end of the workday. This is done to allow the maintenance supervisors time to preview added work, clarify the scope if necessary, and preassign some work for the next workday.
2.4	The planner/scheduler will initiate a "Request for Clearance" for new work orders being added to the daily schedule and requiring such safety clearance. Maintenance supervisors receiving any "Request for Clearance" shall complete the request and submit it to operations before the end of the workday.

Figure 5-7   (*continued*)

workforce levels, the changes adjust the resources available and these changes are in turn reflected in the schedule.

As the planning information for a new work order is entered into the CMMS, the job demands are compared with available resources to determine when the resources will be available. The work order is then scheduled for execution on a specific day. Some of the most sophisticated CMMS programs compare the projected date to the initial priority or requested completion date. If the two dates are in conflict (that is, the work cannot be completed by its requested completion date), the work order is flagged for updating by the planner/scheduler. Either the requested completion date must be changed, or some other work order must be dropped from the schedule on that day.

Key Work Order Scheduling

Many maintenance organizations must consider coordination of crafts in scheduling work. That is, a particular job may require several crafts to complete the work order. This is especially true for organizations with specific craft distinctions. In such an environment, CMMS scheduling programs must be able track and schedule all separate craft work orders that make up the complete job. This is usually done by establishing a *key work order*. This work order is usually the primary craft, or the craft that will complete most of the job. The CMMS scheduling module will schedule this work order first, and then slot all related craft work orders around it.

Limitations to CMMS Scheduling

As they exist today, CMMS scheduling programs can be a powerful tool for the planner/scheduler in preparing work schedules. However, their limitations must be recognized, and often the projected schedule must be modified to fit within the real needs of the facility.

The success of a scheduling system is dependent upon constant adjustments based on changing priorities. These adjustments require good communication with the originators of work, as well as improved knowledge of the facility operation. Failure to update priority data when scheduling with a CMMS is the first step in *invalidating* the system. If changed priorities on work in the backlog are not updated continually, the projected schedules do not reflect the most important work to be performed at any given time.

CMMS scheduling programs currently do little in the area of *optimizing resources*. Schedules might easily have maintenance workers traveling from one side of the facility site to another and then back again in the course of a work day. Common sense would dictate that the maintenance workers would be better deployed without extensive travel time between jobs. This, however, might require the last-minute negotiation of changed priorities or requested completion dates to achieve optimum use of the resources. These adjustments must be made in the field.

Finally, many maintenance backlogs tend to include some amount of *invalid work*. This might consist of duplicate work orders or work that is no longer needed, but never purged from the backlog. CMMS scheduling programs will still recognize these as valid work, and will continue to slot them in the projected schedules. To counter this, a concerted and ongoing discipline must be in place to keep the backlog clean.

Priority Numbering Systems

Many electric utilities and consumer product plants use a priority system that weighs the relative value of each job by using a numerical system. This system applies a numerical weight to each piece of equipment in a facility, and then a numerical weight to each job to be performed on the equipment.

The first step in establishing a priority numbering system is to develop a master equipment list. This list contains all the systems and subsystems in the facility that constitute a maintainable area or equipment.

Equipment List

The name equipment list may be a misnomer because the list usually identifies *locations* in a process or facility. For this reason it may be referred to as a location list, which differs from an asset list. An asset list identifies a specific machine or equipment serial number in the plant and is used mostly for property tax, insurance, or depreciation purposes.

On the other hand, most equipment (or location) lists are originally developed as part of a maintenance management system to capture a history of maintenance expenditures. The level to which a specific location is identified is dependent on the expected return of maintenance history information. For example, a pump in a process may be identified as P-101. Operations personnel may recognize this pump number to include the pump, motor, valves, and associated piping. Facility personnel are only interested in capturing maintenance expenditures on this location, as opposed to only the pump, only the motor, and so on.

Equipment Priority

Next, an *equipment priority number* should be developed for each location. A number from 1 to 9 should be assigned to indicate the relative importance of each location in the process or system. The higher the number assigned to a location, the lower the equipment priority number. Table 5-7 shows an equipment priority system a facility may develop.

Table 5-7   Equipment Priority

Number	Equipment Priority
1	Loss will result in mechanical shutdown of entire facility.
2	Loss will result in the shutdown of more than one process.
3	Loss will result in the shutdown of one process.
4	Loss will result in a significant reduction of capacity.
5	Loss will cut capacity in half.
6	Loss will result in a minor capacity reduction.
7	Loss will result in a slight reduction of capacity.
8	Loss does not affect operation.
9	Nonessential or aesthetic asset.

Facility personnel should review the equipment list and assign an equipment priority number. This task is best assigned to operations personnel, since they are the most knowledgeable of the process or plant operation.

The location number should be assigned whenever a work request is presented to the maintenance department. The equipment priority number should be added to the request by maintenance. This is an automatic process within many CMMSs.

Condition Priority

The originator of the work request would normally describe the nature of the maintenance service required. Additionally, the originator should assign a *condition priority number* from 1 to 8 to the job. Table 5-8 shows common condition priority numbers.

Table 5-8   Condition Priority

Number	Condition Priority
1	A failure has occurred.
2	Failure is likely within the next 24 hours.
3	Preventive or predictive maintenance work.
4	Failure is likely within the week.
5	Failure is likely within a month.
6	Long-range reliability is in jeopardy.
7	Failure is not likely.
8	Modification or minor repair.

Safety or Environmental Modifier

Some facilities may choose to add an additional modifier to equipment and condition priority numbers called a *safety or environmental modifier*. Table 5-9 shows an example of a modifier list.

Overall Priority

All three numbers are multiplied together to generate an overall priority for the job. This gives the safety or environmental modifier significant emphasis because any number multiplied by 0 is 0, which is the highest priority.

Table 5-10 shows an example of a list of work-order priorities.

The first job on the list has a low equipment and condition priority (7 and 6 respectively). However, the zero (0) safety or environmental priority has moved this job to the top of the list. As a matter of fact, most jobs that have a connection to safety or environmental conditions are moved toward the top of the list.

PM work often loses emphasis within many priority systems. This situation is corrected with this priority system. Two PM jobs (condition priority = 3) acquired a relatively high priority.

The originator may still consider adding a *required completion* date to the work request. This would further prioritize the job and can provide a guide to the planner on the best time to schedule the

Table 5-9   Safety or Environmental Priority

Number	Safety or Environmental Priority
0	Immediate threat of injury or exceeds environmental compliance.
1	Potential threat of injury or potentially will exceed environmental compliance.
2	No obvious safety or environmental impact.

Table 5-10   Backlog Priorities

W/O #	Equip #	Equipment Priority	Condition Priority	Safety/ Environment	Priority	Description
566556	37P201	7	6	0	0	Replace guard and nipple.
037911	26T607	1	5	1	5	Weld tank.
432109	47R100	2	3	1	6	Check reactor for leaks at S/D.
550987	47R200	4	2	1	8	Inspect leak at seal.
551806	37R907	2	4	1	8	Replace top rupture disc.
591813	14T375	2	3	2	12	Glass inspection.
102787	05M121	2	6	2	24	Check lubricator on mixer.
541010	37T973	4	8	1	32	Relocate level donuts and dip leg.
468087	05T905	8	5	1	40	New manhole cover for T-905.
504840	10K204	3	8	2	48	Install steam joint.
592772	10P250	5	6	2	60	Reconnect steam tracer.
102808	18C220	8	8	1	64	Install lights.
551447	05T905	7	5	2	70	Replace both sections of rubber hose.
093649	05P102A	8	5	2	80	Install new drip pan.
551808	37R907	6	7	2	84	Remove knock-out pot.
100036	14BLDG	9	8	2	144	Paint "Quality" sign on side.
552780	05P201	9	8	2	144	Drawer is stuck.

job. Also, it may be advantageous to indicate which items are best performed during a facility-wide shutdown.

One disadvantage of a priority numbing system is that it is a static program and does not respond well to changing priorities. Some condition priorities may have to be readjusted as time passes. A job that was assigned a condition priority of 5 (failure is likely in a month) should be modified to a 4 (failure is likely in a week) if the potential problem worsens. This adjustment seldom occurs unless there is a proper forum (such as a scheduling meeting) to renegotiate priorities.

Modifications to Priority Numbering Systems

Large electric utility plants often have two or more operating units that may use different fuels. Fuel prices and overall efficiency for each unit must be factored into the daily priorities. If oil is cheap, a gas unit may be cut back or shut down. All work in that unit should be reduced in importance.

One fix for this situation is to apply a fourth priority number (such as 1 for on-line and 2 for off-line) to all jobs in the backlog. This will tend to move the on-line unit's jobs to the top of the list.

Oil field production is another example of this situation. Some oil wells stop producing for a period of time so their relative importance to the total operation is diminished greatly. When they are producing, the barrels per day or the net revenue value of a well should modify the priority number. A scaling factor, (from, for example, 1 to 9) should be applied to each well based on production.

Additionally, all the oil wells must be put into perspective with other related operations of the oil field (such as dewatering and cogeneration). This is often accomplished by determining an overall weighting factor for each unit. For example, cogeneration plant work requests would be multiplied by 20 percent, dewatering plant requests by 30 percent, and oil well requests by 50 percent.

Summary

No one scheduling method fits every facility, so maintenance scheduling methods tend to vary depending on the type of operation. Whenever any new system is implemented, the basic rules for the effort must be defined. These rules can be the groundwork for identifying the scheduling method that fits best in a facility.

The process used to determine the relative importance of jobs should take into account the long-term and short-term requirements of a facility. This process is called prioritizing. The prioritization process begins as requests for maintenance work are received. It may be helpful to break down the perception of priorities into the three basic stages of classification, requested completion date, and schedule priorities.

The importance of each job on the day it is scheduled can be broken down into two basic categories: planned work (any work in which resource estimating and logistic restrictions dictate or allow the job to be placed on a schedule) and emergency work (any job that displaces work from the schedule). Of the planned work, the classifications of PM, PDM, and prefabrication work have the highest priority, and should be started first. The classifications of corrective and general maintenance work have a lower priority. Emergency work is generally the most expensive kind of maintenance. Emergency work can generally be categorized as real or contrived.

The list of work generated from maintenance requests is called the backlog. This list will grow or shrink as jobs are added or work is completed. Other work (such as emergencies or work that has not been identified with a written work request) is not part of the backlog. As the days progress, a work order in the backlog may not be finished or scheduled in time and, hence, age beyond the requested completion date. The job may no longer be valid, and the originator should remove it from the backlog. On the other hand, the job may still be valid but the priority with respect to the requested completion date may no longer be valid. To ensure that the backlog of work is valid and reflects the actual needs of the facility, steps must be taken to control the backlog.

The allocation scheduling method is most effective in facilities that use the area maintenance concept. This approach refocuses the workforce on PM and PDM work, while providing a facility-wide priority system to complete the other important jobs in the facility. The method lets each operations supervisor or team leader determine the corrective or general maintenance that should be scheduled on a weekly basis. The allocation method is also successful when more than one operations area within the facility is vying for the same maintenance workforce.

All CMMS programs that provide a scheduling module include calendars and resource files. The calendar file is nothing more than a schedule of workdays, and in some of the more sophisticated programs, it also provides information identifying the work hours in a workday. The calendar is interactive with the scheduling module, in that changes within the calendar will immediately be reflected in the schedule. The resource file keeps track of workforce levels, critical support equipment, or other important constraints to any resources that must be tracked. The resource file is also interactive within the scheduling module. As they exist today, CMMS scheduling programs can be a powerful tool for the planner/scheduler in preparing work schedules. However, their limitations must be recognized, and often the projected schedule must be modified to fit within the real needs of the facility.

Many electric utilities and consumer product plants use a priority system that weighs the relative value of each job by using a numerical system. This system applies a numerical weight to each piece of equipment in a facility and then a numerical weight to each job to be performed on the equipment. This system uses an equipment list, an equipment priority, a condition priority, and a safety or environment priority to calculate an overall priority for the job. One disadvantage of a priority numbing system is that it is a static program and does not respond well to changing priorities.

Chapter 6 provides tips and insights for strategically planning a facility shutdown, turnaround, or outage.

Chapter 6

Planning for Shutdowns, Turnarounds, and Outages

Maintenance departments can become a focal point of a company when operations at a plant are halted for some reason. Some plant operations shut down just because inventory is too high, or because business activity may be low. Plants for other companies may be in a sold-out situation, but the companies cannot run their plants in their current conditions. Some organizations (such as the government) may mandate annual shutdowns for inspections. This down period (which may be referred to as a *shutdown,* a *shut-in,* a *downturn,* a *turnaround,* or an *outage*) provides the maintenance department with opportunities that may not arise again for a long period of time. This is a demanding time when the maintenance department must fit a large complement of work into a short period of time.

A shutdown, turnaround, or outage (whatever you choose to call it) is a unique situation that, with proper planning, can afford the maintenance department the window of opportunity to accomplish difficult work under much more accommodating circumstances. Planning a large turnaround, shutdown, or outage can be a planner/scheduler's challenge, or it could prove to be a major disappointment. Shutdown planning can be successful when a meaningful approach is used.

Some companies (mostly utilities) leave outage planning up to one individual or a separate group of people at the facility. Sometimes the planning of an outage at a utility is taken completely away from site personnel and handed over to a corporate engineering department. Facility personnel are queried for their suggestions on the jobs to be performed, but laying out the schedule is left to the specialized group.

This chapter examines how to prepare for a shutdown and how to use some important tools to encourage a successful utilization of this unique opportunity. Several examples help to provide a real-world context to the discussion.

Preparing for a Shutdown

Gearing up for a major shutdown does not necessarily have to be relegated to a special group. A novice with some insight can coordinate a good shutdown. The following steps may be used as a guide.

1. *Job input*—It is imperative that the work covered by a shutdown be clearly defined. This listing of the work also needs a cutoff date. After this date, scheduling and refinement of the execution phase take place.

2. *Shutdown organization*—The entire effort requires an organizational structure specifically assembled to handle the shutdown. This organization's responsibility begins when the first shutdown jobs go to planning, and is not over until the final shutdown report has been submitted.

3. *Execution reporting*—During the actual execution phase of a shutdown, communication is vital. A streamlined reporting and updating vehicle will ensure that work doesn't fall through the cracks. Delays or unforeseen complications are quickly recognized and dealt with. Shutdown status information is most current and real when execution reporting is consistently updated.

A closer look at these three areas will help any plant or facility handle a major shutdown.

Job Input
Defining the scope of the work needed is the first priority of shutdown planning. Input comes from the following sources:

- *Shutdown files*—All maintenance work accumulated against equipment waiting for a shutdown should be reviewed. It is necessary to screen all work in such a holding file to ensure that it is not duplicated through any of the other input sources. It is advisable that the shutdown backlog be printed out, sorted by equipment name or ID, and distributed around to all concerned as a beginning step to the job input phase.

- *Walk-throughs*—Walk-throughs are detailed inspections jointly performed by the operations and maintenance departments. During these walk-throughs, equipment sites are physically visited with the specific intent of identifying any physical problems not previously noticed (or reported). It is advisable for individuals making such inspections to be apprised of identified work in existing shutdown backlogs.

  These walk-throughs are very important. Often, physical problems exist that haven't been reported for months. It is also recommended that operating personnel on shifts be queried, looking for operational anomalies that have been tolerated but must be corrected. It is surprising to discover major equipment problems that have been unreported just because individuals have figured out a way to operate around these problems.

- *Checklists*—All major groups of equipment should have a detailed inspection checklist. Because all facilities have electrical distribution systems, an electrical checklist is included in Appendix A. Similar checklists should be drawn up for other equipment in the plant.

- *PM history*—Although a functional PM effort should have already identified work that must be handled during the shutdown, it is a good idea to review PM history for major equipment. Often, such reviews reveal problems that were not previously recognized.

- *PDM history*—A good PDM program should have identified all equipment that has reached or exceeded engineering limits requiring corrective action. Often, however, pieces of equipment are operating just within acceptable limits prior to a shutdown. This equipment also needs correction during a major shutdown. Otherwise, it will be the first to go over the limit shortly after the plant or facility starts back up. Such immediate maintenance problems after an extensive shutdown undermine the credibility of any maintenance department.

- *Shutdown history*—Shutdown reports after a major effort often single out unique problems noticed during a prior shutdown that should receive further inspection on subsequent shutdowns. Such extra inspections may be outside the normal PM effort. Perusing past shutdown reports will identify any extraordinary work that must be done.

These sources should provide a complete scope of work for the shutdown. After all input is received, the entire list of work should be printed out and sorted by equipment name or ID so that redundancy can be identified. In some cases, work can be grouped together under one work order. The end result is a concise listing of what must be accomplished during the shutdown. After this list is agreed upon, some thought should go into prioritizing the entire list. This initial prioritization will aid in cutting out jobs while scheduling and execution are under way.

Job input must have a cutoff date. This cutoff date is usually a minimum of 2 weeks before the start of the shutdown, but could be as much as 1 month before the shutdown date for a very large shutdown. Only top management can submit shutdown work orders after the cutoff date, and even in this circumstance, some justification must be required.

Shutdown Organization

The shutdown organization is vital to ensure the shutdown is well-planned, all work is covered, and, as the shutdown goes into the execution phase, communication is immediate.

The organization must be headed by one individual—the shutdown manager. All reporting is ultimately given to this individual. The shutdown manager is responsible for guaranteeing that all job inputs are complete. The shutdown manager decides the cutoff date. During the scheduling phase, the shutdown manager decides all scheduling constraints, start and stop dates, and contractor staffing levels.

Reporting to the shutdown manager will be the planning and scheduling group, maintenance engineers responsible for specific jobs, and all shift managers. In turn, the shift managers will each be responsible for a specific shift during the execution. Reporting to the shift managers during execution will be the craft or area supervisors responsible for the work.

It is often a good idea to structure the organization around both *jobs* and *work*. Specific individuals (usually maintenance engineers) will have responsibility over specific jobs. This responsibility covers the scope of work that must be handled on particular equipment. A typical example would be the teardown and overhaul of a large turbine. These individuals keep the shutdown manager apprised of how the job is proceeding. They must notify the manager of problems, percentage of completion, and scheduling projections. Work responsibilities reside with the shift managers. They must keep up with what is happening on a given shift. They must notify the shutdown manager of shift problems, conflicts within load leveling, and schedule compliance on a shift basis.

Execution Reporting

During the actual execution of the shutdown, reporting is essential. Usually this is accomplished through a shift transfer meeting. This meeting is timed an hour before shift change. At the meeting, the shift manager reports on the status of jobs, scheduling problems, leveling conflicts, and completion percentage. The shutdown manager should attempt to attend at least two of these meetings daily. Others in attendance include the shift supervisors, both those at the end of their shift and those at the beginning of their shift, as well as planning personnel.

At this meeting, any changes to the shutdown schedule are considered. Engineers responsible for specific jobs should attend only if there are problems with their jobs that will affect the shutdown schedule. The primary focus of shift transfer meetings is to communicate scheduling problems. These meetings should not be used to handle problems with specific jobs.

The outcome of the meetings will determine any minor changes to the overall shutdown plan. If necessary, shift schedules should be quickly updated so that the oncoming supervisors know exactly what is expected of their shift.

Turnaround Checklist

This section provides items that should be on every shutdown manager's checklist. It consists of items common to almost any shutdown, and each should receive attention when planning for, and dealing with, the logistics of a large shutdown.

Barricades

Barricades should be considered to restrict movement of personnel for any of the following situations:

- To limit entrance to (or egress from) any particular area of the plant or facility.
- To restrict contractor travel to and from the parking lot.
- To protect all personnel from hazardous areas or to minimize access to such areas, and to limit right-to-know training for all temporary personnel.

Building Permits

New construction or major improvements made during a shutdown may require permitting in some locales. Ensuring such legalities are covered in advance of the actual work could eliminate unnecessary and time-consuming delays.

Contractors' Insurance Certificates

Most companies require minimum liability protection as well as proof of worker's compensation insurance coverage for on-site contractors or other outside services. A file should be maintained of these certificates to minimize third-party litigation in the event of injury, death, or major damage.

Dust Control

The extra activity during a large shutdown can also be the source of excessive dust because unpaved areas are often utilized as parking, staging, or even fabrication areas. Contracting a water-truck service to regularly dampen down the areas can keep this problem in check, as well as improve relations with temporary personnel and the quality of work they provide. Providing a temporary wash-down site for automobiles and trucks is also a recommended nicety.

Emergency Showers and Eye Baths
Extra emergency showers and eye baths should always be considered when the number of working personnel increases. These units are available on a rental basis with pressurized water supplies. The rental company can also be contracted to provide regular, documented inspection and testing. You should request copies of such inspections for your own records.

Flag Person or Traffic Control
The use of services or individuals who control traffic or personnel flow should be investigated for large shutdowns. Consideration should be given to covering the following situations:

- Exit to and from temporary parking areas onto local streets during shift changes.
- Traffic control at heavily traveled or centrally located intersections within the plant or facility.
- Special occasions for the movement of heavy machinery, cranes, arrivals of large shipments, or any extraordinary circumstance.

Liquid Waste Handling
Liquid waste from certain cleaning operations may not qualify for handling within the in-plant industrial sewer. These materials must be identified ahead of time for proper handling. If such handling is to be the responsibility of a vendor or contractor, review in detail the method of spill control, containment, and disposal.

Noise Control
Some shutdown operations may generate noise levels that are excessive. These operations must be identified in advance so that proper barricading or posting can be done.

Repairs of Other Damage
All contracts with outside vendors and contractors should include repair clauses for damage to property fences, temporary facilities set up for such personnel, or other plant properties or facilities used by temporary workers.

Repairs of Pavement
Potential damage to pavement areas should be discussed with heavy equipment contractors ahead of time. If load-bearing capacity is unknown, plant roadways should be tested. Contractors should be advised of areas in which damage is probable and should be kept from movement in such areas.

Scaffolding
If several scaffold contractors or rental agencies are to be used at the same time, require that each identify their own scaffolding so that it cannot be mixed up. Requiring a different color from each supplier will help in keeping it all identifiable. During a shutdown, scaffold is often moved from site to site, and the probability of mixing is fairly high.

Solid Waste Handling
As with liquid wastes, potential handling problems can exist for solid wastes, especially when hazardous classifications are involved.

Supervisory Coverage (Dark Shifts and Weekends)
There should always be a company representative on-hand any time temporary personnel (not employees) are in the plant. This individual is responsible for adhering to safety rules and for representing the company in the event of an injury or incident.

Temporary Buildings and Enclosures
Temporary buildings and enclosures are often the direct responsibility and cost of vendors and contractors. It is advisable to review the following areas of coverage with each supplier of temporary structures:

- *Temporary cafeteria or eating facility*—Ensure that some provision is made, including vending equipment. Work through the logistics of restocking vending equipment (when it will be done, which supplier will be used, and so on).
- *Temporary first aid*—Large contractors should provide their own licensed emergency medical technician (EMT) or first-aid technician along with a facility for primary care.
- *Temporary heat and light*—Temporary parking areas used during 24-hour shutdowns should be provided with adequate lighting.
- *Temporary showers and change rooms*—Some shutdown work may necessitate the need for clean and dirty change rooms and shower facilities. The need for (and provision

of) such services should be handled before the shutdown begins.

- *Temporary storage*—Storage for material, tools, and equipment should be the responsibility of the vendor or contractor. Security for such storage and liability if theft or damage occurs should be determined before any material, tools, or equipment come on-site.

- *Temporary telephone*—Temporary telephones should be brought into the plant. These should be located in the normal temporary break areas. It is the responsibility of the vendor or contractor to ensure that abuse of this equipment does not occur.

- *Temporary toilets and water*—Portable toilets and potable water stations should be brought into the temporary structure areas. If these facilities are to be staged within the plant or facility proper, it is advisable to arrange in advance how and when they will be serviced.

- *Temporary power*—If an unusually large contracted workforce is expected, the utilities to such a camp town may tax existing capacity. It's advisable to identify the potential need and provide a temporary source from the local utility.

Temporary Construction Protection

Temporary protection of equipment undergoing maintenance or construction in progress should be addressed. Typical areas to be considered are the following:

- *Large machinery in overhaul*—Protection of bearings, machined surfaces, and tolerance items must be considered.

- *Special attention*—Protection of concrete forms, flange faces of prefabricated piping, and delicate instruments are just a sampling of items that should receive specific attention.

Gang Locks and Shift Locks

OSHA allows the use of *gang locks* and *shift locks* as long as adequate procedures and controls are in place to ensure that such locking devices really provide the necessary protection. It is strongly advised that lock-out procedures be reviewed ahead of time, especially where large numbers of workers are involved, or many different outside companies are on-site at one time.

One of the best tools you can use to schedule a shutdown (large or small) is the critical path method.

Critical Path Method (CPM)

Large jobs such as projects, shutdowns, turnarounds, or outages require a level of organization not normally needed in day-to-day scheduling. Many different (but interdependent) parts of the job must be coordinated. Usually these large jobs extend over days, weeks, and sometimes months. Ensuring that all interrelated tasks are performed on schedule over extended periods of time with many different crews performing the work can be a harrowing task.

During the early days of project management, large scheduling efforts centered on the use of activity bar charts, also referred to as Gantt charts. Gantt charts attempt to display the project's key activities on a time line. Figure 6-1 shows a typical Gantt chart.

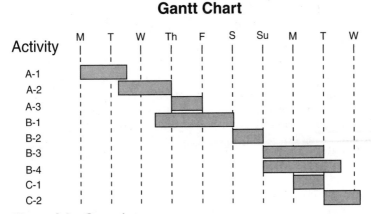

Figure 6-1   Gantt chart.

Gantt charts make it possible to approximate when jobs will start and when they will end. The major deficiency of Gantt charts is the inability to identify project activities that are dependent on each other. In other words, it is not obvious on a large chart that one activity cannot begin until one or more preceding activities are completed.

Interdependencies that are not obvious make it very difficult to identify and react to potential delays on large projects. This is especially true when a key activity is delayed or not completed on time. The Gantt chart must be redrawn or modified to show the new completion time.

Critical path method (CPM) was developed to satisfy this deficiency. CPM is a planning and scheduling technique that was originally developed in conjunction with the building of the Nautilus, the U.S. Navy's first nuclear-powered submarine. The Nautilus project involved more than 200 contract firms. Coordinating the efforts of each phase of the project required that all the relationships between jobs performed by different groups be known at all times. CPM logic diagrams clearly show the dependency of all activities on other activities.

Logic Network Conventions

Critical path networks are represented through either the *arrow diagram method* (ADM) or the *precedent diagram method* (PDM). ADM, the traditional or first method for representing a logic network identifies an *activity* as an *arrow* with *circles* (or *events)*, noting its beginning and ending (see Figure 6-2).

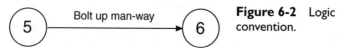

Figure 6-2 Logic convention.

In Figure 6-2, event 5 may denote an event such as "tank is now clean" and event 6 may denote the event "man-way bolted shut." The beginning event is also the ending event for any preceding activities, and the ending event is the beginning event for subsequent activities. The line with the arrow represents the activity "bolt up the man-way." The line and arrow usually point toward the right, but the length of the line does not represent any time or any other resource.

Resources required to perform an activity are often identified on the activity line. Resources are definable inputs that must be allocated to complete activities. Common resources that can be controlled with a CPM logic network are labor, material, supplies, parts, equipment (cranes, lifts, and so on) or time, as shown in Figure 6-3.

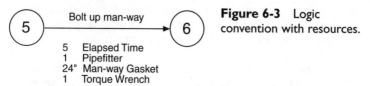

Figure 6-3 Logic convention with resources.

In ADM, the dependency of two activities is identified by the fact that both activities share an event. Suppose that after the activity of bolting-up the man-way is completed, a pressure test can be started. The fact that the pressure test cannot physically be started until the man-way is bolted shut would be diagrammed as shown in Figure 6-4.

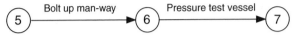

Figure 6-4 Logic convention—activity dependency.

Note that the two activities share event number 6, which is the ending event of the activity of bolting up the man-way, and the beginning event of conducting the pressure test.

The precedent diagramming method (PDM) dispenses with event identification and uses a box to represent the activity. Lines (drawn from left to right) represent the interdependencies of different activities. As with the ADM, these logic lines do not represent any elapsed time. The previous example of two dependent activities can be graphically presented in PDM, as shown in Figure 6-5.

Figure 6-5 Logic convention—precedent diagramming method.

Most computer software programs that perform critical path scheduling graphically portray the network using PDM. (See the section titled *Computerized CPM Programs,* later in this chapter for more on specific programs.)

In either the PDM or ADM diagramming method, the means of determining the dependencies of all activities is the same. Once all activities have been identified and planned (all resources identified), the logic that ties them together is determined by establishing their precedent logic. The activities do not necessarily have to be listed in order. Instead, the proper order is determined by asking the question that establishes the precedent logic for each activity in the list.

Precedent Logic

To prepare a CPM logic network, the discreet activities are identified first. They are then placed in order. Any activity's place in the order is determined by the activity's dependency on:

- Any preceding activities
- Any activities that follow the activity
- Activities that can go on concurrently

It is not necessary to be this rigorous in building the logic network. Viewing every job on the list with these three categories in mind is a formidable task.

The logic network can be built by asking only one question of each activity.

Which activity or activities immediately precede this activity?

Answering this question for every activity on the list automatically determines the sequence and interrelation of activities. This method of putting activities in order is called *precedent logic*.

Now let's take a look at a simple example.

Salami Sandwich Project

Assume we have the following activities (with associated times for completion):

A. Spread mayo on bottom slice of bread—25 seconds.

B. Lay salami on mayo side of bottom slice—10 seconds.

C. Spread butter on top slice of bread—25 seconds.

D. Spread mustard on buttered side of top slice—25 seconds.

E. Lay top slice of bread on salami (mustard down)—10 seconds.

The first step is to determine the precedent activities. Ask the following question for each activity: "Which activity or activities immediately precede this activity?" The result will be as follows:

- Activity A has no precedents.
- Activity A is a precedent to activity B.
- Activity C has no precedent.
- Activity C is a precedent to activity D.
- Activity B and D are precedent to activity E.

The complete network is then diagramed as shown in Figure 6-6.

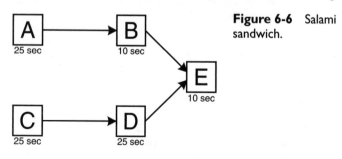

Figure 6-6 Salami sandwich.

It's not a good idea to put constraints on the resources when developing a logic diagram. The assumption is made that unlimited resources exist. The fact that one slice of bread could have been prepared concurrently with the other slice may not have been evident at first. If resources were available (for example, another pair of hands), you could build the sandwich in a shorter period of time.

Finding the Critical Path

CPM scheduling is usually used to determine the shortest amount of elapsed time required to complete a series of interdependent activities. There are usually several paths of related activities that will start at the beginning event of the first activity and continue until the ending event of the last activity. There will usually be one path, however, that will require the most elapsed time from the start to the end.

In our sandwich project, activities A, B, and E will take 45 seconds to complete, but activities C, D, and E will take 60 seconds. Activities C, D, and E define the shortest period of time in which the whole job can be completed (as long as two people make the sandwich). This path is known as the *critical path*.

Earliest Completion Time, Latest Completion Time, and Float

Besides identifying the critical path, other vital information can be found within the network diagram. This information comes from determining the *earliest completion time* and *latest completion time* of the activities in the logic network. By definition, earliest completion time is the calculation of the earliest possible finish time of each activity. Earliest completion times are calculated by starting at the beginning event and adding up subsequent activity times. Similarly, the latest completion times are the latest possible times that each activity can be finished without increasing the length of the project. They are determined by starting with the total project elapsed time at the end of the network and subtracting activity times until the first activity is reached.

With the salami sandwich project, the bottom slice of bread with the mayo is not critical. In fact, there could be 4 extra seconds for anyone working on the bottom slice if the work began at the same time as the work on the top slice. This difference is called *float* (or *slack*).

Float can be more technically defined as the *difference between the earliest completion time and latest completion time* for each activity, as shown here:

$$F = T_L - T_E$$

In our sandwich project, the earliest completion time for the bottom slice (activities A and B) was 35 seconds into the job. The preparation of that slice, however, did not have to be completed until 50 seconds had elapsed (the latest completion time). The shared float for activities A and B is 15 seconds, as shown here:

$$F = T_L - T_E = 50 \text{ sec.} - 35 \text{ sec.} = 15 \text{ sec.}$$

The earliest and latest completion times for the activities defining the critical path were equal. The difference between the two times (float) for all those activities is 0. Another, more technical, definition of the critical path is a series of activities in a network with no float.

Float can be used to develop a priority list for all activities not associated with the critical path. Activities with little float will be put near the top of the priority list, and activities with a lot of float will be near the bottom of the list. Also, by moving the start time of an activity with float, the planner can level a resource load. This process will be explained later under load leveling.

Prioritizing work based on available float is also helpful in highlighting *off-site work*. These activities can take on an out of sight, out of mind appearance, especially if they have some float. Material that is on order, or equipment that is out for repair, may not be in the critical path. However, if the float time is minimal, any delay in the delivery can quickly make those activities part of a new critical path. If the float is exceeded, the completion time of the project will be extended and a new critical path will develop.

Now let's take a look at another example.

Piping System Replacement Project

In this example, we will first identify the problem, and then devise a solution using the principles we have learned about the critical path tool.

The Problem

A thickness check of critical piping has detected a 150-foot section of piping in an overhead rack that requires immediate replacement. This particular section of line operates at an elevated temperature and is heavily insulated. All the valves installed in the pipe section should also be replaced. There is no extra room in the pipe rack, so the existing 150-foot section must be removed before the new section can be installed.

An inspection of the current engineering files reveals that the latest drawings do not reflect an as-built condition. Therefore, scaffolding will have to be erected and actual flange-to-flange dimensions measured before replacement spools can be fabricated. Also, product inventories are low, so downtime while the line is being replaced must be kept to an absolute minimum.

The Solution

It is decided that a critical path schedule is necessary to minimize plant downtime. While thinking through the job steps, several innovative ideas are considered. The planner decides to prefabricate piping subassemblies in the shop. A rough material list (including pipe lengths and bends) can be quickly developed by visual inventory. The fabrication work can start as soon as the pipe and fittings have been received. When the piping subassemblies are completed, they can be moved to the site where the replacement spool sections can be field-welded. Accurate piping spool dimensions can be obtained after the scaffolding is in place. As the project is developed, it becomes apparent that a pressure test must be performed after the piping is installed and before the pipeline is insulated.

Table 6-1 shows the list of activities. Notice that they are not in order.

The first step in developing the logic network is to put the activities in order. This is best done using the precedent logic method described previously. The activities are listed again in Table 6-2. The activity or activities that must precede each activity are identified in the far-left column. This step simplifies construction of the final

Table 6-1   Activity List

Item	Activity	Activity Duration (hours)
A	Pressure test new line.	6
B	Remove existing pipeline and insulation.	35
C	Purchase and receive valves (9 days).	216
D	Remove scaffolding and clean up debris.	4
E	Insulate new line.	24
F	Move valves from warehouse to job site.	2
G	Develop material list.	8
H	Erect scaffold.	12
I	Purchase and receive pipe (8 days).	192
J	Prefabricate pipe subassemblies in shop.	40
K	Deactivate old pipe line (operations).	8
L	Move pipe subassemblies to the field.	8
M	Field-weld subassemblies into replacement sections.	16
N	Install new pipe and valves.	8

Table 6-2   Precedent Logic

Item	Activity	Precedent
A	Pressure test new line.	N
B	Remove existing pipeline and insulation.	H, K
C	Purchase and receive valves (9 days).	G
D	Remove scaffolding and clean up debris.	E
E	Insulate new line.	A
F	Move valves from warehouse to job site.	C
G	Develop material list.	—
H	Erect scaffold.	—
I	Purchase and receive pipe (8 days).	G
J	Prefabricate pipe subassemblies in shop.	I
K	Deactivate old pipe line (operations).	—
L	Move pipe subassemblies to the field.	J
M	Field-weld subassemblies into replacement sections.	L
N	Install new pipe and valves.	B, F & M

diagram. Using this table, a completed network diagram can be drawn as shown in Figure 6-7.

The total time to complete all the activities in the example project could be as much as 579 hours (the duration of all task estimates added together). However, some activities can be performed simultaneously. This is something that may not have been evident prior to the development of the logic network. Additionally, the critical path of activities is also identified. The list activities (in the center of this diagram) constitute the longest path in the project (activities G, I, J, L, M, N, A, E, and D). The total time required by the project, assuming sufficient resources are available, is only 306 hours.

The critical path activities have no available float time (as per the definition). All other activities in the network can be delayed or extended by a calculated amount of float. For example, activity C (purchase and receive valves), has 38 hours of float (262 − 224 = 38 hours). This means that the vendor has only 38 hours of leeway on a 9-day estimate to provide the valves. It would be prudent to closely follow this activity since it can become critical if it is delayed more than 38 hours.

Activity K (deactivating the old pipeline) is an operations activity. Operational prejudices may compel the planner to assume that other activities (such as the receipt of all materials) would be precedent

Piping Project

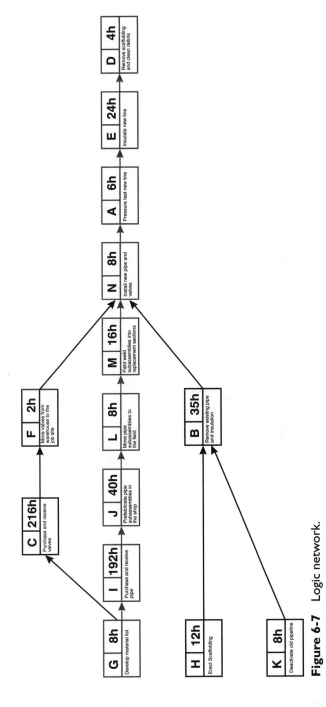

Figure 6-7 Logic network.

to this activity. However, there is nothing physically preventing the pipeline from being shut down at any time. Adherence to the pure logic helps us discover another possibility. The total float for activity K is 221 hours ($229 - 8 = 221$ hours), which means operations can continue to operate for 221 hours while other activities are being performed.

Since reducing downtime is so important, the planner should also review duration estimates for the list of jobs that require downtime. Activities K, B, N, and A make up this list. The total downtime can conceivably be 81 hours. Activity B (removing the existing line—35 hours) makes up a large part of this downtime. The planner/scheduler may decide to increase the number of pipefitters assigned to this task to reduce the downtime.

Time Domain Logic Network
A *time domain logic network* combines the critical path method advantages with the ease of viewing a project in Gantt chart form. Many project-management software programs offer time domain networks as an option. In a time domain network, activities are plotted to show their elapsed time, as well as the dependencies with prior and later activities, as shown in Figure 6-8.

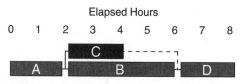

Figure 6-8 Gantt chart and critical jobs.

Activities A, B, and D are the critical path with a project duration of 8 hours. Activity C is noncritical, with a float or slack of 2 hours.

Reducing Project Time
One of the goals of a project may be to minimize the elapsed time for each activity. As stated previously, by developing a CPM logic network, concurrent activities will become evident. A planner can use this information to staff different activities with different crews working simultaneously to efficiently use the workforce.

Another time-saver may be to staff-up on some activities. In other words, if a particular activity can be completed in half the time by doubling up on the resources, then the elapsed time will decrease. This concept has limitations. The law of diminishing return stipulates that increasing some resources will not always result in completing the activity in a shorter time. There are often space restrictions on the number or people who can realistically be assigned to any given activity. If limited downtime is the goal of a particular project, some inefficiencies in job staffing can be tolerated. However, resources added to shorten the activity could be equal to throwing money away.

Project Duration versus Project Cost

Often, a planner is faced with an economic decision during a project plan. A balance may have to be made that minimizes both the project cost and the project duration. To determine the true relationship between project cost and project duration, three parameters must be determined:

- *Direct project cost*—The direct project cost is all the money spent on the resources in the project. Project costs and duration for different resource conditions should be calculated and plotted.

- *Downtime cost*—A project may require a shutdown of the facility. The cumulative cost of downtime must be considered if this is the case.

- *Crash time*—There is a maximum number of resources that will actually result in the shortest project duration. The shortest duration is called the crash time of the project.

Figure 6-9 shows the relationship of the project cost, downtime cost, and the project crash time. The *total cost curve* is the sum of the project cost under different resource conditions and the downtime cost under increasing duration of the project. Notice that the

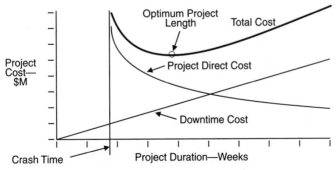

Figure 6-9 Project cost and downtime.

optimum duration for the project is not the minimum time in which the project can be completed.

The crash time must be considered for each activity on the critical path. There are two good reasons to do this:

- *Determining the crash time provides important management input*—Many managers not familiar with maintenance are of the opinion that throwing additional resources at a project *has* to reduce its duration, which may not be the case for some jobs.

- *Operation demands can change*—Upper management in some companies may change their minds on how long of a shutdown can be tolerated. Sometimes this happens in the middle of the shutdown. Having determined the crash time for the total project and each activity on the critical path, maintenance is in a good position to update management on the options.

Load Leveling and Project Constraints

The logic network, by itself, does not always produce the optimum plan. Often, the plan that is developed uses the facility's labor resources in an inefficient manner. Figure 6-10 shows what can happen with one craft in a facility.

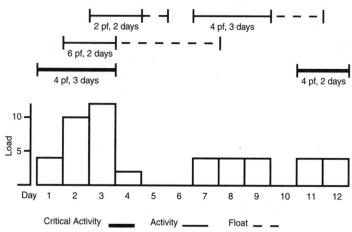

Figure 6-10 Load leveling 1.

This chart represents activities of only one part of the total project plan. There are five jobs for the pipefitter craft during the outage, two of which are on the critical path. A problem arises from the fact that the facility has only four pipefitters. The project seems to require eight pipefitters on the second day and 10 on the third. The noncritical jobs have some float, so the planner decides to *level the load* on the pipefitter craft.

Load leveling is the distribution of resources so that constraints on the resources are not violated. The planner decides to use the float to *change the start date* on all noncritical jobs. This, in itself, does not balance the load. The planner then looks at *increasing or decreasing the staffing* on some of the jobs, including the critical path jobs, to balance the load. Figure 6-11 shows the resulting plan.

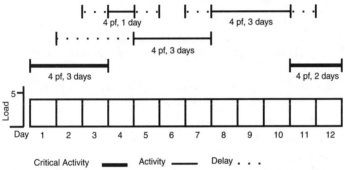

Figure 6-11 Load leveling 2.

It is important to note that although the pipefitter craft is now balanced, some activities are delayed until the last moment. These activities no longer have any float. These activities may be part of a path including other jobs with no float. This, by definition, is the new critical path. The project duration may be extended under these conditions, so the planner will need to redraw the critical path on a time line. The portion of the float that is consumed by the delay of an activity must now be considered part of the activity duration.

CPM computer programs make the task of load leveling a simple one. The planner need only identify the resources and constraints on each resource. The program will then recalculate and display the resource load throughout the project. On request, the program can adjust load to use up the float. If conflicts still exist, the planner can increase or decrease resources assigned to some activities to further balance the load. (Even though some programs can also make staffing adjustments on activities as part of load leveling, this work is better left to the planner to avoid unreasonable over- or under-staffing of some jobs.)

Other resources that can be constrained are outside services (usually large and special cranes). At the beginning of a large shutdown, many lifts may be required. They must be coordinated with available cranes. One problem that occurs often is placing cranes too close to each other. A large crane making a lift may block the movement of another crane. In shutdown planning, all these considerations must be taken into account by the planner. These skills are obviously beyond the ability of a computer program.

Computerized CPM Programs

For those planners/schedulers who will be using CPM methods repeatedly, a PC-based project scheduling program should be considered. The advantages of these programs are as follows:

* Resource levels can be established ahead of time. The advantage here is that resources will not be violated without your knowledge.
* Most have a built-in calendar module that allows you to establish ahead of time work hours and work days.
* Most are built around precedent logic.
* The critical path is highlighted.
* Changes to the project are easy to make and a new logic diagram is easy to print.

The following list of project software is not a complete list, but it is a good sample of programs that can be used in *maintenance project scheduling.*

Following are some less-expensive project programs:

* *Microsoft Project*—Microsoft Corp., One Microsoft Way, Redmond, WA 98052-6399, (800) 426-9400, (425) 882-8080, www.microsoft.com/project.
* *Project Scheduler*—Scitor Corp., 333 Middlefield Rd., 2nd Floor, Menlo Park, CA 94025, (800) 533-9876, (650) 462-4200, www.scitor.com.
* *SureTrak Project Manager*—Primavera Systems Inc., Two Bala Plaza, Bala Cynwyd, PA 19004, (610) 667-8600, (800) 423-0245, www.primavera.com.

Following are some higher-end project programs:

* *Primavera Project Planner (P3) for Windows*—Primavera Systems Inc., Two Bala Plaza, Bala Cynwyd, PA 19004, (610) 667-8600, (800) 423-0245, www.primavera.com.

- *Open Plan Professional for Windows*—Welcom Software Technology, 15995 N Barkers Landing, Suite 275, Houston, TX 77079, (800) 274-4978, www.wst.com.
- *Artemis Views*—Artemis Corp., Boulder, CO and Burlington, MA, (800) 477-6648, www.artemispm.com.

Shutdown Tips

Finally, here are some tips that have helped many other planners/schedulers:

- After load leveling is performed, work lists are generated. A work list itemizes all activities, sorting them by craft or crew. Work lists are easier to work from than critical paths.
- Make up a blind list. Know where every blind is installed, and make sure that each is removed.
- Put up all scaffolding beforehand.
- Never schedule two jobs in the same place at the same time. It sounds crazy, but it can happen.
- String temporary lighting beforehand. *Test it out!*
- Set up smoking areas and portable toilets ahead of time.
- If cranes are scheduled, watch out that jobs aren't scheduled under the crane's lift path.
- Have all possible materials that can be delivered in advance located at the job site.
- Have a dry run of the shutdown. Walk everybody through it.
- Decide in advance how new jobs will be handled.

Summary

A shutdown, turnaround, or outage is a unique situation that, with proper planning, can afford the maintenance department the window of opportunity to accomplish difficult work under much more accommodating circumstances. Gearing up for a major shutdown does not necessarily have to be relegated to a special group. A novice with some insight can coordinate a good shutdown by defining job input, the shutdown organization, and execution reporting.

Defining the scope of the work needed is the first priority of shutdown planning. Job input comes from shutdown files, walkthroughs, checklists, PM history, PDM history, and shutdown history. These sources should provide a complete scope of work for the shutdown. After all input is received, the entire list of work should be printed out and sorted by equipment name or ID so that redundancy can be identified.

The shutdown organization is vital to ensure the shutdown is well-planned, all work is covered, and as the shutdown goes into the execution phase, communication is immediate. The organization must be headed by one individual—the shutdown manager. All reporting is ultimately done to this individual. The shutdown manager is responsible for guaranteeing that all job inputs are complete.

During the actual execution of the shutdown, reporting is essential. Usually this is accomplished through a shift transfer meeting. This meeting is timed an hour before shift change. At the meeting, the shift manager reports on the status of jobs, scheduling problems, leveling conflicts, and completion percentage. At this meeting, any changes to the shutdown schedule are considered. The primary focus of the shift transfer meetings is to communicate scheduling problems. These meetings should not be used to handle problems with specific jobs. The outcome of the meeting will determine any minor changes to the overall shutdown plan.

The critical path method (CPM) was developed to satisfy deficiencies related to Gantt charts. Critical path networks are represented through either the arrow diagram method (ADM) or the precedent diagram method (PDM). ADM, the traditional or first method for representing a logic network, identifies an activity as an arrow with circles (or events), noting its beginning and ending. PDM dispenses with event identification and uses a box to represent the activity. Lines (drawn from left to right) represent the interdependencies of different activities.

In either the PDM or ADM diagramming method, the means of determining the dependencies of all activities is the same. Once all activities have been identified and planned (all resources identified), the logic that ties them together is determined by establishing their precedent logic. The activities do not necessarily have to be listed in order. Instead, the proper order is determined by asking the question that establishes the precedent logic for each activity in the list.

CPM scheduling is usually used to determine the shortest amount of elapsed time required to complete a series of interdependent activities. There are usually several paths of related activities that will start at the beginning event of the first activity and continue until the ending event of the last activity. There will usually be one path, however, that will require the most elapsed time from the start to the end.

A time domain logic network combines the critical path method advantages with the ease of viewing a project in Gantt chart form.

Many project-management software programs offer time domain networks as an option. In a time domain network, activities are plotted to show their elapsed time, as well as the dependencies with prior and later activities.

Load leveling is the distribution of resources so that constraints on the resources are not violated. CPM computer programs make the task of load leveling a simple one. The planner need only identify the resources and constraints on each resource. The program will then recalculate and display the resource load throughout the project.

For those planners/schedulers who will be using CPM methods repeatedly, a PC-based project scheduling program should be considered.

Chapter 7 explores how to gather important data to be used when compiling maintenance performance indices.

Chapter 7

Gathering Data for Maintenance Performance Indices

The purpose of an index is to help predict future activity or to compare current activity to a standard. Indices can help identify negative trends before they become too costly. They can also bring to light the success or failure of programs recently instituted by management. Planners and managers have used indices to help justify changes in the maintenance department or just toot their own horn.

The value of an index is related to the cost of obtaining it. If the index requires that data be gathered by maintenance hourly employees, it is probably not worth the effort. Maintenance workers are usually paid much too much for this type of data gathering. If the data for the index does not flow naturally from normal maintenance record keeping, don't develop new procedures to get the information (such as forms, cards, and time clocks). A good rule of thumb is if the data for an index takes more than 5 minutes per week to collect, it most likely is not worth the effort.

The *process* of collecting the data should be set up one time, and the task of developing the data should be moved down to the clerical level if possible. The data should be collected and manipulated on a weekly basis.

Often, some extra detail must be added to a work order form as well as the data entry screen for a computer database to facilitate sorting of the data. Plant area and equipment number are the most common criteria used. Entry questions and codes indicating the work class (PM, corrective, general, and so on) or priority (planned versus emergency) are also common in work-order systems.

Sometimes the list of codes required to be entered into the system can be very large. When the code list is long, or too many parameters are requested, a default code is often chosen, or the parameter requirement is ignored completely. The amount of codes to be entered by an originator, planner, or supervisor should be kept down to an absolute minimum to improve the chances of it being entered.

Data is much easier to compile and manipulate if it is all in the same place. It is desirable that 100 percent of the work performed by a maintenance organization be charged to a work order, as opposed to an area or piece of equipment. This will improve the data-collection process. This means a separate work order must be written

for each job performed. Blanket or standing work orders are sometimes employed to capture the hours worked on short jobs. Standing
work orders should not be overused. Charges to these work orders
should be limited to less than 10 percent of the total work order
charges.

The data can be collected manually or by computer. Standard, off-
the-shelf maintenance management programs often do not provide
the exact data in the specific form required. It is always far easier to
manipulate the data if it is down-loaded into a spreadsheet program
(such as Lotus 1-2-3, Quattro, or Excel). Data can be downloaded
from *any* computer system to a format that can be imported by these
programs. A list of work-order record fields to be downloaded from
a CMMS should include the following:

- Work order number
- Priority
- Date initiated
- Required completion date
- Short description of job
- Actual hours (completed work)
- Estimated hours (backlog work)
- Craft
- Equipment number
- Area
- Work classification (PM, corrective, and so on)
- Material dollar charged to work order
- Originator

The following pages describe sources of data used to calculate maintenance indices.

Completed Work Data

Consider the following when gathering data related to completed
work:

- Total completed labor hours
- Emergency work orders and labor hours
- PM labor hours completed
- Overtime

Total Completed Labor Hours

The total labor hours are not the total paid hours, as is often assumed to be the case. Actually, the total paid hours are made up of three costs:

- Hours charged to work orders
- Hours worked but not charged to work orders
- Hours paid but not worked

The hours worked but not charged to work orders can never be captured. The hours paid but not worked account for the time paid for meetings, training, and other nonwork activity. Both of these values are usually minimal in size; no time should be spent tabulating them. Attempts to account for all paid hours usually require a cumbersome record-keeping process with little payback.

The hours charged to work orders provide the best accounting for the total labor hours. Actual hours worked should be indicated on every work order before it is closed out. This data is best retrieved by tabulating data from completed work orders. A weekly tabulation of the completed work orders and the hours charged to the work orders is the best way to keep up with the record keeping.

Breakdowns of the completed hour data by craft, area, or equipment can be very helpful when analyzing the data.

Emergency Work Orders and Labor Hours

This is a subset of the total work completed in a week, month, or year. Usually a separate (emergency) priority is marked on the work order, which makes it easier to sort through manually or by computer.

In some facilities, certain emergency work is not normally charged to a work order. Work such as a megohm check of a motor to see if it has failed is sometimes charged to a standing work order. First, the hours charged to standing work orders must be totaled. Next, an *estimate* must be made of percentage emergency hours charged to standing work orders. This part of the standing work order hours can then be added to the hours already tabulated for emergency work orders. This portion of the emergency work should be ignored if it is less than 10 percent of the total emergency work.

PM Labor Hours Completed

PM work is another subset of the total work completed at a facility. These work orders should be identified as PM to make them easier to sort.

It is normal for a minimal amount of corrective work to be charged to PM work orders instead of writing a separate work order. If an abnormal amount of corrective work is charged to PM work orders, it may have to be manually extracted, because it tends to artificially increase the actual PM hours. Some companies use the estimated hours instead of actual hours for completed work to get around this problem. This is an acceptable practice if PM work estimates closely approximate actual hours.

Information on PM data, broken out by craft, area, or equipment, is very useful when analyzing the data. A subsort by these categories should also be considered. Work that was performed on many pieces of equipment at the same time (such as lubrication routes) may not be easy to separate by equipment. The hours charged to these types of work orders can be tabulated under an "other" or lubrication category. The hours charged could also be manually prorated for each piece of equipment on which the work was performed.

Overtime

Overtime may not be indicated on the completed work order. Usually, only the actual hours are entered, without any indication regarding what part of the number is overtime. It may be necessary to derive overtime data from payroll data. Time cards or time record sheets indicate the hours paid for overtime. This number should not be compared to the total paid hours, but rather should be used for a close approximation of the overtime on work orders. As a normal course, these hours should be compared to total completed hours charged to work orders.

Overtime can be subtracted from the total completed hours charged to work orders to determine a value for *straight time*. Here again, payroll data will not provide the best indication of straight time, since some of the time paid was not really worked.

Backlog Data

Consider the following when gathering data related to backlogs:

- Backlog hours
- Planner's backlog
- Crew week

Backlog Hours

The backlog is the total estimated hours of work waiting to be completed. Maintenance backlog can include several different elements. Jobs that have been planned, estimated, and are ready to work are definitely part of the backlog. It is also a good idea to add jobs that have been estimated but are waiting for parts, as long as the parts are slated to arrive before the scheduled start date of the job.

Backlog work should be separated by craft and by area. The craft data can be used to calculate weeks of backlog. The area data may help in determining future staffing requirements under an area maintenance concept.

It may also be a good idea to break backlog data out by *class of work*. Adjustments must often be made in a backlog calculation for shutdown or PM work. Other routine work that is covered on a work order generated every week should also be tabulated.

Planner's Backlog
Jobs that have not yet been estimated or planned cannot be considered part of the backlog of work, but rather as the *planner's backlog*. A count of jobs that are open but have no estimate is all that would be required here.

Crew Week
The number of labor hours available on any given week is used in backlog week calculations and in plotting the backlog. In its most basic form, the crew week is the number of maintenance workers times the number of hours in a normal workweek. Usually it is sufficient to know just the average work force available in a year times the normal workweek.

Schedule Data
Consider the following when gathering data related to schedules:

- Schedule hours
- Actual hours of scheduled work completed
- Number of days in the schedule

Scheduled Hours
The best way to collect scheduled hours is to manually total all hours scheduled during each week or month from printed work schedules. Longer jobs that are scheduled to be completed over more than one week should be accounted for by prorating the completed portion of job. A status report on every job should be provided by the first-line maintenance supervisor to the maintenance planner on a daily basis, so the process of updating the status of long jobs will be automatic.

Over the course of time (such as a month), the effect of longer jobs on the scheduled hour total will be minimal. Therefore, it may be unnecessary to make adjustments for longer jobs if the scheduled hours are totaled monthly rather than weekly.

Actual Hours of Scheduled Work Completed

The actual hours worked on a schedule during a week or month should be totaled. This number can quickly be derived by updating all daily schedules with the actual hours and totaling both actual and estimated hours. Other jobs (such as emergency work and add-on work) should not be included in the actual value. Some CMMSs make this easier to do through an ad hoc report.

Number of Days on the Schedule

Many times, jobs are placed on a schedule and are not started or completed. It is important to keep track of this activity to determine the success or failure of the scheduling process. Some computer programs keep track of the number of times a job is put on a schedule. If this software feature is not available, a manual record must be made (preferably by a clerk) while preparing the daily schedules.

Base Cost Data

Consider the following when gathering data related to base costs:

* Cost of the maintenance hour
* Total maintenance costs
* Cost by equipment
* Total contract costs
* Cost of production
* Book value of facility

Cost of the Maintenance Hour

It is a good practice to estimate the dollar value of a work order even if it is not a requirement of the facility. Estimating the cost of a work order can cast resource-consuming jobs into a new light. A modification may be deemed unwise when it is found that it will cost the facility $10,000 or $100,000 to complete.

The hourly pay for a maintenance worker is not the only cost associated with the labor. Other items, such as vacation pay, holiday pay, and benefits are part of the cost of having the worker on the payroll. Estimates for maintenance dollars should include all costs associated with that individual. This combined rate is called the *standard labor rate*. A standard labor rate can be between 1.3 to 2 times the hourly labor rate, depending on the benefit package.

Standard cost systems are often used in industry for internal performance measurement and control. They provide the best estimate of how well maintenance budget goals are being met.

The following annual costs are often considered when developing a standard labor rate for maintenance workers.

- *Item 1*—Hourly labor rate × 8 hours × # of employees × # of work days
- *Item 2*—Overtime percent (last year) × 1.5 × Item 1
- *Item 3*—Sick pay
- *Item 4*—Holiday pay
- *Item 5*—Vacation pay
- *Item 6*—Other nonwork pay
- *Item 7*—Benefit costs (includes company contributions to FICA)

The following example explains how a standard labor rate may be established.

Example—Assumptions

A plant has 100 hourly maintenance employees. The pay rate is $15.00 per hour. Each employee should work about 235 days next year. No pay is made for sick time. Overtime amounted to 11 percent of the hours worked last year. The company gives 10 paid holidays per year. The total vacation projected for all employees is 1500 days. Each employee is paid $10 per week for a clothing allowance. The total benefit package amounts to $1.3 million per year.

Example—Calculations

Using these assumptions, the following annual costs can be calculated:

- *Item 1*—$15/hr. × 8 hours × 100 employees × 235 days = $2,820,000
- *Item 2*—11 percent × 1.5 × $2,820,000 = $465,300
- *Item 3*—N/A
- *Item 4*—10 holidays × 8 hours × $15/hr. × 100 employees = $120,000
- *Item 5*—1500 days × 8 hours × $15/hr. = $180,000
- *Item 6*—$10/week × 50 weeks × 100 employees = $50,000
- *Item 7*—$1,300,000

- *Total Items 1–7—$4,935,300*
- *Effective Working Days*—235 days + 235 days × 11 percent = 261 days

Thus, the following formula calculates the hourly wage:

$$\text{Labor Rate} = \frac{\$4,935,300}{261 \text{ days} \times 8 \text{ hours} \times 100 \text{ employees}}$$

$$= \$24 \text{ per hour}$$

All dollar estimates of maintenance work at this plant should be based on a labor rate of $24.00 per hour. This rate will be accurate for the coming year. Most companies update the standard labor rate whenever the labor contract changes or when the annual budget is developed.

Some maintenance organizations (mostly building maintenance) include supervisor, planner, and other support personnel salaries in the standard labor rate calculation. Other facilities charge one rate for day-to-day maintenance work, and a higher rate (including support salaries) for capital projects.

Total Maintenance Costs

On an annual basis, the accounts payable and payroll records should produce the most exact figure of total maintenance costs. Month to month, these costs should be derived from payroll hours recorded during the month times a standard rate. The standard rate should include the hourly pay, benefits, and overtime.

Payroll records can also be used to add in the *salaries* of management and support employees in the maintenance department. Here again, benefits and overtime pay may have to be included.

Parts and material charges can be determined from direct purchases of material by the maintenance department. This becomes a slightly tricky exercise when parts are bought for storage rather than for immediate use. The way this data is extracted depends on the way it will be used. Material or parts that are purchased for storage might not be used within the period in question.

Items purchased for a storeroom can either be expensed at the time of purchase, or charged to an inventory account and only expensed at the time of use. For example, if all parts purchased for storage were expensed at the time of purchase, the cost of these purchases over a period could be totaled and then added to the other material purchases. Over the course of a year, this type of material cost record keeping would most likely approximate actual use. However, month-to-month tracking of these values may lead to erroneous conclusions. This is because heavy inventory purchases during a single month may be much higher or lower than the *use* of the material during that month.

Month-to-month tracking will provide a more accurate accounting of maintenance material activity if most of the stores items are only expensed at the time of use. Facilities that operate in this way charge the maintenance budget only when parts are withdrawn from the storeroom.

Cost by Equipment
Maintenance costs are very helpful if they can be isolated by equipment or area. Data that include labor and material charged to equipment can provide some insight into the bad actors or equipment that cost the most for the facility to maintain.

Direct purchases of material used by maintenance should be charged to a work order. In other words, a work order must be generated for every maintenance purchase. The work order can be traced to a piece of equipment by an equipment number. If this data is stored on the same CMMS as all other data, the effort of collecting costs is simplified. If this is not the case, the data must be combined manually.

Total Contract Costs
Contract costs should be tabulated on a monthly basis. Here again, all contract costs are easier to track if they are charged to a work order. Accounts payable data may have to be reviewed to get an exact cost number.

Cost of Production
Operations should be consulted on the cost of production. This information will include the cost of raw material and the salaries of production employees. The data is most useful when it is provided by line or process train. Most production cost information is used to evaluate the cost of downtime.

Book Value of Facility
It can be helpful to know the book value of a facility when comparing maintenance costs from one site to another. The *book value* is the purchase and construction cost of buildings and equipment in a facility, less the total depreciation of those assets.

Companies that are publicly traded are required by the Securities Exchange Commission (SEC) to publish asset information in their annual reports or 10K supplements. Such assets are usually broken down into four categories:

- Buildings, machinery, and equipment
- Real estate and general facilities
- Construction in progress
- Mineral and energy reserves

Top management will often compare maintenance costs between companies by indexing those costs to historical asset dollars. Depending on the mix of assets, such comparisons can be misleading. It is better to compare company-wide maintenance costs to only the maintainable assets, such as the first category in the previous list.

Maintenance and repair costs expressed as a percentage of assets are also affected by capital investment and equipment upgrades. When capital replacement is curtailed, maintenance costs increase on old equipment.

Other Data
Consider the following when gathering other data:

- Downtime hours
- Equipment availability
- Inadequate repairs
- Corrective action work derived from the PM effort
- Equipment reliability data

Downtime Hours
Most facilities are not very diligent when it comes to recording downtime. This is mainly attributable to functional reasons rather than an attempt to hide something. As with maintenance data, downtime data is only worth the effort if there is a reasonable return and it doesn't take too long to collect. The return is judged to be minimal by most organizations because the data is never used when it is collected. Additionally, operations personnel feel their time is better spent making improvements to the operation than recording the time it is down.

When it is recorded, an operations logbook is the only source of a downtime record. Data may be entered by a shift employee to inform the next shift of problems encountered and the current status of the operation.

Some companies use sophisticated production software to record downtime activity. Maintenance employees usually have little direct access to this data, and it usually is not connected to the CMMS. The causes of downtime can be categorized in the following manner:

- Equipment failure (working down inventory)
- Equipment failure (no inventory)
- Scheduled down
- Maintenance work (portion of failure time)
- Process problems (not equipment-related)

When equipment in a process or production chain fails in service, an organization has a few choices. It can issue a request for maintenance work and then proceed to satisfy finished product requirements by working down inventory. Operators on the line will have to be reassigned if possible. No charge should be made against maintenance for downtime under these conditions until the inventory is used up.

If an operation is in a sold-out situation, there is no inventory to work down. Under these circumstances, the maintenance department will be indirectly responsible for the downtime. They are directly responsible when the inventory is gone. When they release the equipment back to operations, it may not be immediately returned to service because of operations scheduling problems. Maintenance should not be charged with this downtime.

Errors in record keeping are common when downtime is recorded. One common mistake occurs when the equipment is initially marked as down for maintenance, but the maintenance employee determines that it is really a process problem. Maintenance management must be diligent in the correction of these kinds of record-keeping errors if the downtime data is going to have any meaning at all.

Equipment Availability
This is basically the opposite of downtime. It is calculated by subtracting downtime from the hours the equipment is scheduled for operation.

Inadequate Repairs
Some candid maintenance managers admit that many of the repairs that are performed on a day-to-day basis are repeat jobs. Unfortunately, a large part of these repeat jobs are caused by faulty maintenance repairs. Obtaining data to substantiate this claim can be a cumbersome, manual process. However, the data may help the maintenance manager justify a larger training budget.

Corrective Action Work Derived from the PM Effort

It does no one any good to perform PM work that provides little benefit in the long run. It is more helpful to determine how effective the PM program is. One aspect of good PM work is the ability to identify problems that would otherwise turn into an emergency.

Tabulating the number of corrective-action work orders derived from PM provides one true measurement of a PM program. One rule of thumb commonly used by maintenance managers is that at least three corrective action jobs should be derived from every 10 PM jobs performed. Tabulating corrective work derived from PM may have to be performed during a scheduling meeting.

Equipment Reliability Data

Condition-based monitoring provides the best insight into the reliability of the equipment. Measurements such as vibration and lubrication analysis not only provide insight into problems, but can also help classify the types of problems.

Summary

The purpose of an index is to help predict future activity or to compare current activity to a standard. Indices can help identify negative trends before they become too costly. They can also bring to light the success or failure of programs recently instituted by management. The value of an index is related to the cost of obtaining it. If the index requires that data be gathered by maintenance hourly employees, it is probably not worth the effort. Maintenance workers are usually paid much too much for this type of data gathering. If the data for the index does not flow naturally from normal maintenance record keeping, don't develop new procedures to get the information (such as forms, cards, and time clocks). A good rule of thumb is if the data for an index takes more than 5 minutes per week to collect, it most likely is not worth the effort.

Data is much easier to compile and manipulate if it is all in the same place. It is desirable that 100 percent of the work performed by a maintenance organization be charged to a work order, as opposed to an area or piece of equipment. This will improve the data-collection process. This means a separate work order must be written for each job performed. Blanket or standing work orders are sometimes employed to capture the hours worked on short jobs. Standing work orders should not be overused. Charges to these work orders should be limited to less than 10 percent of the total work order charges.

Data collected for indexing completed work include total completed labor hours, emergency work orders and labor hours, PM labor hours completed, and overtime.

Data collected for indexing backlogs include backlog hours, planner's backlog, and crew week.

Data collected for indexing scheduling activities include scheduled hours, actual hours of scheduled work completed, and number of days on the schedule.

Data collected for indexing base costs include cost of the maintenance hour, total maintenance costs, cost by equipment, total contract costs, cost of production, and book value of a facility.

Other data that may be collected include downtime hours, equipment availability, inadequate repairs, corrective action work derived from the PM effort, and equipment reliability data.

Chapter 8 shows how to transform this indexing data into indices that can be used to measure performance.

Chapter 8

Using Indices to Measure Performance

Professional credibility is achieved when you can prove the truth behind your statements with supporting data. Sometimes, managers and planners/schedulers may be put in the position of having to defend their goals, decisions, or current situation. While these professionals may feel as though their reputations and experience should provide credibility for their statements, that credibility can be damaged without supporting source documentation.

Chapter 7 provided some guidelines for collecting raw data for maintenance performance indices. Presentation of the data should have some traceability. This means that managers should be able to track their information from the source documents through their analysis and conclusions. Statements and numbers found in the source documents are logged into a common record such as the CMMS. The raw and unaltered numbers are used to generate indices on the performance of the planning and scheduling effort.

Developing raw data into a useful index requires some talent, experimentation, and patience. Indices must be reviewed over time to see whether they provide insight into the operation of the maintenance department. An index that proves to be very useful at one kind of facility will be useless at another.

This chapter provides insights on developing effective indices for several areas, including the following:

- Backlogs
- Schedule compliance
- Estimating accuracy
- Emergency and preventive maintenance
- Overtime
- Productivity

We begin the discussion by examining a sample scenario in which the manager was actually required to provide more than one index.

The Need for More Than One Index

Whenever an index is used in a report, it is important to present the information truthfully. This usually requires that more than one index be presented to show a trend or to indicate an interrelation between indices.

Interrelationships between indices result from the constraints imposed on the maintenance workforce. There are usually a fixed number of technicians to perform the work. There are a minimum number of hours a day a technician can work efficiently. If the reliability of the equipment is low, the emergency rate may be high. In a 24-hour operation, a high rate of emergencies usually requires longer working hours for the technicians (that is, overtime).

The following example shows how indices can be deceptive, depending on how they are presented. A maintenance manager has been asked by the plant manager to provide some statistics on the effectiveness of the maintenance effort to reduce emergencies. The maintenance manager provided the memo shown in Figure 8-1.

	InterOffice Correspondence
DATE:	3/10/03
FROM:	Maintenance Manager
TO:	Plant Manager
RE:	Monthly Report

Here are the monthly maintenance statistics you asked for in the staff meeting. I think we're doing well.

% Emergencies (Labor hours)	
January 2003	14%
February 2003	11%

The Maintenance Manager

Figure 8-1   Short-sighted memo.

The plant manager is impressed with the reduction in emergencies and commends the maintenance manager on the perceived efforts in this area. However, the plant manager is also worried that overtime may be getting out of hand and asks the maintenance manager to add this data before resubmitting the report. The memo shown in Figure 8-2 was resubmitted by the maintenance manager.

The plant manager was still pleased with the emergency rate, but the fear about overtime seems to be a fact. Sensing that there is some sort of relationship between the two figures, the plant manager asks for the details. The last memo the maintenance manager provided is shown in Figure 8-3.

The detail provided in the last memo shows the possibility that PM work (a figure not requested by the plant manager) is being

	InterOffice Correspondence	
DATE:	3/12/03	
FROM:	Maintenance Manager	
TO:	Plant Manager	
RE:	Monthly Report	

Here again are the monthly maintenance statistics you asked for in the staff meeting. Call me if you have any questions.

	% Emergencies (Labor hours)	% Overtime (Labor hours)
January 2003	14%	14%
February 2003	11%	33%

The Maintenance Manager

Figure 8-2   Insightful memo.

performed on overtime. This may or may not be desirable to the plant manager. The fact that the original emergency rate figures reported by the maintenance manager were distorted by the overtime rate does concern the plant manager.

Now let's take a look as some specific types of indices used to measure performance.

	InterOffice Correspondence
DATE:	3/14/03
FROM:	Maintenance Manager
TO:	Plant Manager
RE:	Monthly Report

Here are the revised monthly maintenance statistics you asked for, with a little more detail. I'd like to talk to you about this data, if I could. I don't think it's as bad as it looks.

January 2003

$$\% \ Es = \frac{Es}{Es + Normal + PM} = \frac{500}{500 + 2000 + 1000} = 14\%$$

$$\% \ OT = \frac{OT}{OT + ST} = \frac{500}{500 + 3000} = 14\%$$

February 2003

$$\% \ Es = \frac{Es}{Es + Normal + PM} = \frac{500}{500 + 2000 + 2000} = 11\%$$

$$\% \ OT = \frac{OT}{OT + ST} = \frac{1500}{1500 + 3000} = 33\%$$

The Ex-Maintenance Manager

Figure 8-3   Seeing 20/20.

Backlog Indices

A backlog index can be an indicator of the ability of the maintenance department to keep up with the new work that is generated every day. It can also be used to determine the proper size of the workforce. The backlog of maintenance work is the total estimated labor hours for all planned work. An index can be developed that gives one number to the backlog, expressed in *weeks*, using the following formula:

Backlog =

$$\frac{\text{Backlog Hours}}{\text{Hours in a Work Week} \times \text{Number of Available Craftspeople}}$$

The *workable backlog* index is calculated from a total of all the work that can be done immediately because parts and other materials are on-site. The *total backlog* is calculated from the work that is estimated and planned, as well as parts and materials that are on order or are available now. The workable backlog should be a minimum of 2 weeks, and the total backlog should be a minimum of 4 weeks. These limits hold true for most facilities. The upper limit varies with each maintenance organization and craft type.

Most backlogs should be maintained within the range of 2 to 8 weeks. It becomes very hard to put together an efficient schedule if the backlog is low. Even if the work is estimated and all the parts are available, operations may not be ready to release the equipment for repair.

One cause of a low backlog is late identification of needed repairs. If a large portion of maintenance work is identified at the last minute, the backlog may remain low. The solution to this problem is an improved PM program coupled with attempts to sensitize operations to potential equipment problems.

If the backlog is consistently low and the emergency rate is also low, the workforce may be too big. An example of this situation is a facility that may have increased its staff for the start-up of a new plant. Emergencies may be commonplace while the plant is in the early stages of start-up. As the plant matures, the emergency rate drops. A cut-back in head count is one solution to this situation. Another may be to have the maintenance department start taking on capital projects rather than hiring a contractor.

If the backlog is high, higher-priority work may not be getting done on time. Temporary increases in overtime or the use of contract labor may cure this problem in the short run. If the backlog is consistently high, an increase in the workforce may be necessary. This increase should occur only after a detailed review of the work in the backlog. If a large portion of the backlog is just fill-in work, personnel changes are unwise. This work should be purged from the backlog, because it is unnecessary to the operation of the facility. If the high backlog is caused by capital work for a construction campaign, most likely the maintenance workforce should not be used to perform this work and contractors should be hired.

Maintenance productivity improvements (such as detailed job plans and closer coordination with operations) are also ways to reduce the backlog. If an employee's workday can be made more efficient, more work can be performed and the backlog should go down.

Many backlogs are listed by area to help adjust workforce requirements from one area to the next. In plants with more than one craft, backlog indices are more usable when they are calculated for each craft. Some crafts may be able to sustain higher backlogs than other crafts. This fact can be determined over time by closely monitoring the index and trying to relate changes in the index with scheduling problems, as shown in Figure 8-4.

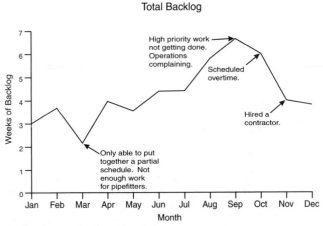

Figure 8-4   Backlog for 1 year.

The backlog dropped to about 2 weeks in March. The planner noted that the pipefitting backlog during March was not sufficient to support the current staff of pipefitters.

The facility went on a campaign to increase the backlog by identifying needed work in the facility that had been previously overlooked. This seemed to work, but the backlog eventually grew to a point that left more and more high-priority work uncompleted in September. Operations personnel complained that important work was not getting done on time, creating increased emergencies and downtime. To combat these problems, the maintenance manager called for scheduled overtime to work down the backlog. This seemed to work at first, but high-priority work was still not getting done in October. The maintenance manager hired a contractor to work down some of the backlog.

All of the reactions described here lagged the problem. As a result, most of the corrections turned out to be too little, too late. Looking at the backlog data with attendance data helps reveal one source of the problem, as shown in Figure 8-5.

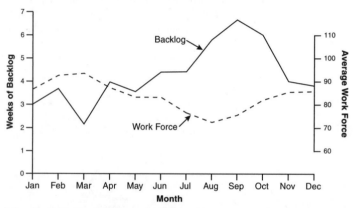

Backlog and Work Force

Figure 8-5   Backlog and attendance for 1 year.

Attendance seems to be at least one of the factors affecting a change in backlog. The backlog was at its lowest in March when attendance was at its highest. The backlog went up in August and September after attendance started to drop (because of vacations). The effect of attendance on the backlog seems to be delayed by about a month, so plotting attendance may help the planner predict backlog levels.

Higher backlogs in June and July could be worked down with scheduled overtime and contract help. The backlog remaining in the lean months, August and September, should then be much lower than the previous year. The goal of these adjustments will be a relatively level backlog, varying between 3 and 5½ weeks.

PM work can often confuse the calculation of backlog. PM work is usually performed at an even rate throughout the year. As a result, a certain percentage of the workforce will be consistently committed to PM work. This percentage should be *backed out* of the workforce for the backlog calculation. This problem is further complicated by the fact that PM work may already be part of the backlog; some of this work may be overdue PM, and should be left in as valid backlog. Other PM work may just have been dropped into the backlog by the CMMS. The backlog calculation is often modified by backing out the labor hours associated with this PM work.

Consider the following example to show how this adjustment might be made. There are 96 line maintenance employees (excluding stores). The normal workweek is 40 hours long. The schedulable work in the backlog is 22,100 hours. PM work currently in the backlog is 4300 hours, with 25 percent of the completed work being PM. So, the PM workforce can be calculated as follows:

$$0.25 \times 96 = 24 \text{ employees}$$

$$\text{Backlog} = \frac{22{,}100 \text{ hours} - 4300 \text{ hours}}{(96 \text{ employees} - 24 \text{ employees}) \times 40 \text{ hours/week}}$$

$$= 6.2 \text{ weeks}$$

With PM work backed out of the calculation, it is apparent that sufficient workforce exists to perform PM work and a 6.2 week backlog exists for the remaining workforce.

Shutdown, outage, or *turnaround* work in the backlog must also be taken into account. No adjustment is made if the shutdown work in the backlog is just for a short outage that does not require additional outside labor. If the shutdown work in the backlog is earmarked for a major shutdown (which will be staffed through scheduled overtime and contract assistance), then this work should be removed from the backlog before a calculation is made.

Backlog Age

Priorities assigned to work orders can be expressed in terms of allowable days, weeks, or months before a job must be completed. *Backlog age* is a measurement of the number of work orders in the backlog that can be completed within their priority period and the number that have missed the deadline. It is sometimes expressed as a percentage of the total number of work orders in the backlog that have aged beyond their priorities.

Backlog age is best charted to show the actual number of work orders for each priority period. Table 8-1 shows an example of this method.

Table 8-1 Work Order Age

| | | Priority | | | |
Age	E—Immediately	R1—Within 48 Hours	R2—Within 1 Week	R3—Within 1 Month	R4—Over 1 Month
1 day	6	21	37	22	17
2 days	2	13	18	12	15
3 days to 1 week	1	6	25	16	12
1 week to 1 month	0	3	14	18	16
Greater than 1 month	0	0	9	29	17

All the work orders in the shaded areas of the table are over-aged. For the example, out of 329 work orders in the backlog, 64 have aged beyond their priority limit.

An over-aged backlog is indicative of problems originators have with the priority system and the maintenance department's capability to meet the originator's requirements. This measurement also provides a reading of the overall credibility of the maintenance effort. One such credibility problem shows up when the majority of the backlog is high priority (E or R1 on our example). An originator who really requires a job to be completed within a month, but places a high priority indicating it should be performed sooner, does so because the originator has little confidence that the work will be completed on time. If work cannot be completed within the priority period, the originator should be notified, and a new priority should be negotiated. A continuous process of updating the priorities is less obtrusive than one-shot reprioritization of the backlog. Planning meetings are the best forums for altering priorities.

Ideally, no work orders should age beyond their priority period. In actuality, this is impractical in even the best-operated maintenance department. To maintain credibility with operations or any other originator, an attempt should be made to keep over-aged work orders from exceeding 10 percent of the total work orders.

Schedule Compliance

Schedule compliance is the value representing maintenance jobs actually completed according to the established schedule. This is usually expressed as a percentage of hours scheduled. The following equation shows this relationship:

$$\text{Schedule Compliance} = \frac{\text{Scheduled Hours Actually Worked}}{\text{Total Hours Scheduled}} \times 100\%$$

Ideally, schedule compliance would be 100 percent. In the real world, where schedule breaks and emergencies occur, 90-percent schedule compliance can be an achievable goal. If work is constantly displaced by emergency work or last-minute additions, the maintenance schedule is a superfluous document. On the other hand, if the schedule compliance is 100 percent and emergency work and add-ons are also performed, then the labor estimates for scheduled work may be inflated. Schedule compliance is one index that can be plotted and compared to a plot of emergencies, as shown in Figure 8-6.

In Figure 8-6, there seems to be an inverse relationship between *schedule compliance* and emergencies. This would be understandable. True emergency work should displace scheduled jobs. Added

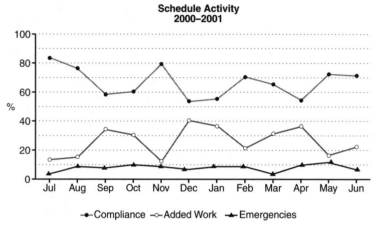

Figure 8-6   Schedule compliance and emergencies.

work, also shown on the graph, bears no relationship to schedule compliance. This is work that was added during the week but was not emergency work. In other words, the work is important enough to be completed within a few days of initiation. The commitment to completing a weekly schedule is put into question if the added work ever becomes a larger part of the work completed during a month.

Estimating Accuracy

Estimating accuracy is determined by dividing the *actual* hours of completed work by the *estimated* hours. Emergency work and other jobs that must be estimated on the fly (because of a perceived urgency) should not be included in this calculation. Estimating accuracy is often expressed as a percentage of the total actual work to total estimated work completed over a set period. Estimating accuracy should fall within the range of 85 percent to 115 percent.

Another way to present this data is with a *scatter diagram*. A scatter diagram is a method of graphing data to show the actual distribution of the data. Figure 8-7 shows an example of actual-versus-estimated hours.

The hours estimated by the planner were based on sound methods, so most of the estimates were not multiples of 4 or 8 hours. Each mark (×) indicates a point described by the estimated and actual hours for one work order. The diagonal line across the center of the graph defines the point where actual hours equal estimated hours. All marks above the line are jobs that took longer than estimated,

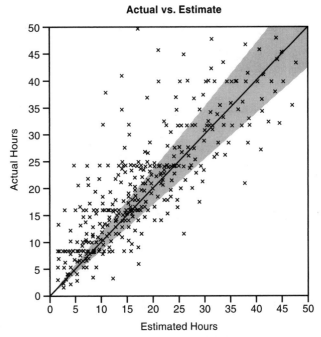

Figure 8-7 Scatter diagram of actual versus estimated.

and all jobs below the line took less time than estimated. The shaded area describes the planner's goal of ±15 percent of the estimate.

There are some interesting things to point out about this particular diagram. First, it's obvious that the majority of the actual hours are greater than estimated. This can be an indication of a few things:

- The planner is estimating based on a best-case scenario and may not have taken into account the problems commonly encountered in maintaining this facility.
- The planner may not be allowing for a large enough crew size.
- The planner is estimating jobs correctly, but inefficiencies exist in the workforce that make the jobs take longer.
- The supervisors may be overstaffing some jobs.

Upon further inspection, you can see that many jobs took a set amount of labor hours to complete (that is, 8, 16, 24, 32, or 40 hours). This is evidenced by the grouping of marks in a horizontal direction at these time periods. This usually occurs when a job is finished within the last 2 hours of an 8-hour day. Some supervisors may not have any more work for the maintenance employees in their charge at the end of the day, so they are assigned to sweep the shop or do some other minor work. The extra time is charged to the work order they started in the morning.

The scatter diagram can be very helpful to a maintenance manager when attempting to evaluate the planning process. Problems such as the ones described here are just a few of the possible situations that can exist. A manager should also become suspicious if the actual and estimate are consistently the same. This may mean the planner is a very good estimator. However, very few plans really meet the estimate, so some other reason must exist for this perceived accuracy. Sometimes, when the estimated hours are shown on the maintenance employees work order, the employee will take that estimated time to perform the work, even if less time is actually required. As the job begins to wind down, the employee may pace himself or herself to meet the estimate. A supervisor should be alerted to this situation.

Evaluating Actual versus Estimate

Actual versus estimate comparisons *can* be used by a maintenance manager as one indicator of the success of the planning program. Some of the issues to examine are:

- Possible overstaffing
- Bad labor estimates
- Lack of parts
- Equipment still running
- Poor work performance

Possible Overstaffing

If the actual consistently exceeds the estimate, the supervisor or maintenance workers may be padding timesheets to make up for time not really worked. This is often a problem if the workforce is too large for the workload. This starts with the planner/scheduler issuing a lean schedule, which does not fill up the work day of each maintenance employee. The maintenance employee or supervisor performs the work on the schedule. If very few emergencies or add-on jobs come up, the maintenance employee stretches out the work to be performed or just hides until the day is over.

The planner/scheduler should be instructed to fill up the schedule with work. If not enough work is available for the workforce, the maintenance manager should consider a reduction in the workforce.

Bad Labor Estimates

The planner could also be the source of poor actual-versus-estimate performance. Steps in the plan may have been missed or addition errors could have occurred. These problems are excusable if they occur infrequently. If the estimate was a *guess,* and not a judgment, then the planner should definitely be faulted. The manager should investigate further to be sure that proper time was available to perform a decent plan.

Lack of Parts

Lack of parts is one of the common reasons maintenance supervisors give for job extension. Again, this is the responsibility of the planner unless there is no way the need for certain parts could have been predicted in advance.

Equipment Still Running

Many jobs performed by maintenance require the equipment to be shut down. The planner should arrange for sufficient downtime with operations before putting the job on the schedule. If operations forgets to comply or consistently has little regard for the time lost by maintenance in these circumstances, the maintenance manager should discuss the situation with the operations managers.

Poor Work Performance

Some managers attempt to use actual versus estimated hour indices as *productivity indices,* assuming the estimate is correct and the actual is incorrect if they are different. Productivity is best measured directly, through *supervision.* Other managers attempt to use standard job estimates to identify poor performers in the department. This is a questionable form of management and long discredited. The supervisor reviews the average time to complete certain jobs and compares this average to the performance of individuals in the department. The supervisor may use any difference on the low side as a justification to discipline, and even terminate, an employee.

Poor performance usually has many roots or may not exist at all. First of all, the work of the slower person may be of a higher quality than that of the faster employees. Secondly, the employee may not be properly trained for the job at hand, or the employee may not be properly motivated to perform at a higher level. These problems require management and leadership on the part of the supervisor and seldom require disciplinary action.

The best time for the maintenance manager to investigate these discrepancies is at the end of each day. An update meeting with the planners and supervisors provides the best forum to get to the bottom of the problem. Asking a simple question—why wasn't this job finished?—will usually start the ball rolling. Filtering the truth from the lies is what the manager is paid to do.

PM and Emergency Indices

The easiest PM indicator is the percentage of PM work completed over a given period. If PM accounts for most of the completed work at a facility, emergency work is usually minimal. If little or no PM is performed, emergency work will be commonplace. Most industrial facilities have a running PM percentage between 20 percent and 40 percent.

PM work is sometimes put off if other, more pressing work comes along. Most facilities find that this excursion is paid for in the long run. PM work should be some of the highest priority work on a schedule. Plotting PM work over time can indicate the declining or improving ability of a facility to complete PM work, as shown in Figure 8-8.

Figure 8-8 shows the PM percent (the bars) for each month in the year and indicates the emergency percent for these months. The

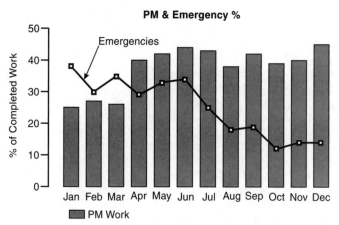

Figure 8-8 Preventive maintenance and emergencies.

facility decided to beef up the PM effort and to make a concerted attempt to get that work done. The effect on emergencies was not immediate, but over time, the emergency percentage started to fall. This lag was mainly caused by a delay in completing corrective action work orders derived from PM inspections.

Major swings in the emergency rate make it difficult to evaluate trends. A good way to smooth out the peaks, and to gather some insight into the real emergency situation, employs a *moving average,* as shown in Figure 8-9.

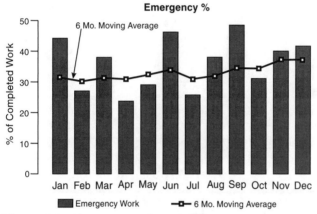

Figure 8-9   Emergencies plotted with moving average.

Figure 8-9 shows an emergency rate that jumps up and down from month to month. These wide swings may coincide with production changes, but they tend to hide the *true emergency rate* of the facility. Plotting a moving average tends to smooth out the peaks. A *moving average* is the average of data in a window over a set period (such as 3 months, 6 months, or 1 year). For example, a 6-month average is calculated by adding all the data for the month in question and the 5 months prior, and then dividing it by six. So, for the month of June, the moving average would be as follows:

$$\text{Moving Average for June} = \frac{\text{Jan} + \text{Feb} + \text{Mar} + \text{Apr} + \text{May} + \text{Jun}}{6}$$

The moving average in Figure 8-9 reveals a trend upward in the emergency rate, from about 30 percent to about 35 percent, which would not have been noticed otherwise.

Other PM Indicators

Another important PM indicator is the age of PM work in the backlog. As stated previously, PM work should be included in every weekly schedule. It should also be completed on time. If a frequency is reasonably set, the PM should be completed the week it comes due. Emergencies occasionally occur that may supersede PM work for one day, but this work should reschedule for the next day.

A simple count of the PM work orders not completed in the prior week is a good indicator of the effectiveness of the PM effort. A goal should be set not only to complete the overdue PM work orders, but also to complete every PM work order in the week it is due.

Another PM indicator is a count of the corrective-action repairs or work orders derived from PM. This index was explained in Chapter 7.

Overtime

The overtime percentage should be calculated and plotted on a weekly basis. A 3-month moving average may be helpful when it comes to identifying longer-term trends, as shown in Figure 8-10.

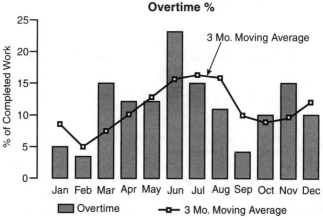

Figure 8-10 Overtime plotted with moving average.

Keeping overtime down to a minimum is vital in managing a budget. However, occasional overtime is unavoidable. Some managers claim that working overtime is actually cheaper than hiring new employees, even at the premium rate of time and a half. This is because most of the benefits for the employees working overtime are already paid for during the regular hours. This is true for many benefits, but it is not the case for the company contribution to FICA, which is based on the total earnings.

Additionally, overtime work is not necessarily as productive as work performed during the normal workday. First, there is no supervision, so good work performance may be a question. Secondly, the employee has already worked a full day and may be tired during the overtime period. Mistakes caused by hurrying through the job (or inattentiveness) will quickly eat into the savings perceived by overtime.

Some overtime work can be planned, such as overtime used to work down a backlog or catch up on PM work. However, overtime work often goes hand-in-hand with emergency work, as shown in Figure 8-11.

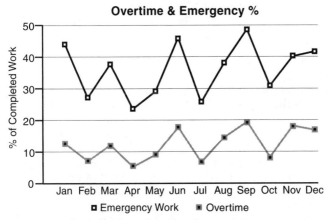

Figure 8-11   Overtime plotted with emergency percentage.

Figure 8-11 shows overtime plotted with emergency work. As the emergency rate increases, so does the overtime rate. Most of this overtime is obviously not planned. The problem here is not overtime, but rather excessive emergencies.

Productivity Indicators

Productivity of the maintenance workforce is easier to control than it is to measure. Very few indicators exist that truly show the productivity of the workforce. Some common indices used are the following:

- Ratio of maintenance labor costs to maintenance material costs
- Maintenance cost per unit of production
- Actual maintenance cost compared to budget costs

Ratio of Maintenance Labor Costs to Maintenance Material Costs

Basically, this index is an attempt to determine how efficiently the maintenance material dollars are being used by the maintenance workforce. This index can be confused by an increase in labor costs caused by labor contract increases. An increasing index could be indicative of higher productivity if the labor rate and workforce stayed the same.

Studying this index from one month to the next provides little insight into the activity of the maintenance department. This figure is best trended on an annual basis.

Maintenance Cost per Unit of Production

Calculating the cost of maintenance per pound, ton, or other unit of production can be helpful when pricing a product or determining profitability of the product. However, this figure may provide little information as to the productivity of the workforce. When production goes down at some facilities maintenance costs go up. This is because the maintenance department is working on the equipment that is down, *building capacity* back into it. On the other hand, if production goes up, maintenance costs may stay the same. Prefabrication work may be performed for coming outages, shutdowns, and turnarounds, and may occupy the maintenance workforce.

This index is best trended from year-to-year, rather than month-to-month.

Actual Maintenance Cost Compared to Budget Costs

Too many managers track actual maintenance costs against the budget with little idea of how to correct negative trends. If the actual monthly costs are below the budget, they're happy and decide to spend more next month. If the actual costs are higher than budget, they don't know what to do. This is usually because the maintenance budget is laid out in *line accounts,* which are helpful to the accounting department but provide little insight for the maintenance manager.

Table 8-2   Plant Goals

Plant Goals	% of Completed Work
Corrective Work	31%
PM Work	45%
Shutdown Work	22%
Emergencies	2%

Meeting the maintenance budget should be a goal for the maintenance department. If the budget was laid out properly, an estimate of PM, corrective, shutdown, and emergency work should have been developed. The first step to meeting the budget is to see how well the goals set for these activities are met. Table 8-2 shows an example of some budget goals for these items.

With these goals in mind, changes in month-to-month cost data will make much more sense. For example, assume emergencies went up 25 percent over the goal during the month. If this coincides with an increase in the total actual costs for the month, the maintenance manager should look more closely at the cause of the failures during that month. This may lead to an investigation of the effectiveness of the PM effort.

A Discredited Index

The utilization factor is the actual hours charged to work orders compared to paid hours. The utilization factor is said to be one indicator of maintenance productivity. Meetings and paid-cleanup make up the nonwork part of the percentage. There is little benefit to be gained by this index, because it provides no additional insight into how to correct the problem, if there really is one.

Managing with Indices

Although maintenance indices are not meant to relieve the maintenance manager from actual hands-on management of assigned areas of responsibility, those indices can point to areas on which attention must be focused.

One index by itself will rarely identify an area of concern. Several indices should be evaluated to qualify a situation. An increase in emergencies might be the cause of reduced PM. Higher emergencies may also result in an increase in backlog. This is normally because the priority system has broken down.

Indices are of minimal value when analyzed at a single point in time. Rather, all indices should be compared to past values. In analyzing any indices, trends can identify if a critical situation is developing, improving, stabilizing, or getting worse.

Case Study: What's the Problem Here?

An individual recently relocated to a new site to become the maintenance manager. The graphs shown in Figure 8-12 indicate key costs and several performance indices for the mechanical department. With the exception of the new manager, the only departmental personnel changes had been a new mechanical supervisor hired in March of 1992. What could be the reason for the increasing maintenance costs?

An analysis of the data reveals some inconsistencies. In the last year and 9 months, there has been an increase in both schedule compliance and completed PM, but there has also been a steady increase in emergencies since the middle of 1992. If real PM is being performed, emergencies should not be on the increase, but rather would be reduced even further.

The improvement in schedule compliance is inconsistent with the increased emergencies. Usually, emergencies displace scheduled work. As emergencies increase, schedule compliance decreases.

Some answers to these inconsistencies are revealed (or at least suggested) by the data alone. The mechanical maintenance cost information reveals that the labor and material costs are relatively stable. The cost that is increasing is in contracted maintenance. It can be inferred that the increased emergencies are being handled with contracted maintenance.

It should finally be noticed that these inconsistencies coincide with the hiring of the new mechanical supervisor. At this point the maintenance manager should investigate further into several areas of concern. Does the increase in completed PM represent valid work, or is the supervisor allowing PM work orders to be pencil-whipped? Why are contracted maintenance personnel being used to perform emergency repairs? Is the supervisor being outwitted by the workforce? Has supervisor control been lost in this case?

Basic Maintenance Reporting

It's a good idea for the maintenance department to develop a weekly report of the activities. The report should consist of a *balance sheet, performance indices,* and *graphical presentation.* The indices, tables, charts, and graphs shown in Chapter 7 can be a part of the report, but a balance sheet should always be included.

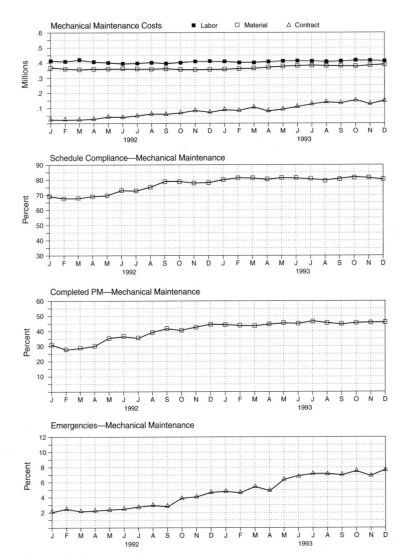

Figure 8-12 Case study.

Backlog Data:

	# of W/Os	Est. Hrs.	Act. Hrs.
Backlog as of 12/3/03	1223	13842	
Requests Received and Est.	280	2177	
Less Completed Work Orders	(322)	(1927)	(1969)
Net on Hand	1181		
43 Work Orders in Progress:			
Completed Hours		(156)	155
Remaining Hours		161	
Total Completed Work			2124 Hrs
Backlog as of 12/10/03	1181	14,097 Hrs	

Backlog Craft Analysis:

Craft	# of W/Os	Est. Hrs.	Weeks
Electrician (8)	214	1856	5.8
Instrument (4)	92	747	4.7
Pipefitter (10)	275	3205	8.0
Welder (11)	245	3344	7.6
Millwright (17)	214	2992	4.4
Contractor	141	1953	–
	1181	14,097	6.1 Avg. (Non-Cntr.)

Backlog Status Analysis:

	# of W/Os	Est. Hrs.
Emergency	2	9
Planned	571	8821
PM/PDM	247	1147
Shutdown	361	4120
	1181	14,097

Completed Work Analysis:

	# of W/Os	Est. Hrs.	Act. Hrs.
Emergency	53	332	332
Planned	199	1517	1592

Figure 8-13 Backlog data.

Balance Sheet

The balance sheet is an accounting of the raw data used to generate performance indices and graphic presentations in the report. It's a look at the maintenance numbers at a point in time and is equivalent to balancing a checkbook or chart of accounts for a business. Figure 8-13 shows an example of this part of the report.

The data at the bottom of this balance sheet will be used at the top of next week's report. An analysis or further breakdown of the backlog can be added to the balance sheet (Figure 8-14 and Figure 8-15). The totals found on the balance sheet should agree with the totals found in this data.

Backlog Craft Analysis:

Craft	# of W/Os	Est. Hrs.	Weeks
Electrician (8)	214	1856	5.8
Instrument (4)	92	747	4.7
Pipe Fitter (10)	275	3205	8.0
Welder (11)	245	3344	7.6
Millwright (17)	214	2992	4.4
Contractor	141	1953	–
	1181	14,097	6.1 Avg. (Non-Cntr.)

Figure 8-14   Backlog craft analysis.

Backlog Status Analysis:

	# of W/Os	Est. Hrs.
Emergency	2	9
Planned	571	8821
PM/PDM	247	1147
Shutdown	361	4120
	1181	14,097

Figure 8-15   Backlog status analysis.

Analysis of the completed work can also be provided, as shown in Figure 8-16.

Performance Indices

In addition to the balance sheet section of the report, a section of *performance indices* should be presented, as shown in Figure 8-17.

The weekly report can, of course, be modified from the form described here. It may also be advantageous to develop a *monthly report* with additional data as well as graphics.

Completed Work Analysis:

	# of W/Os	Est. Hrs.	Act. Hrs.
Emergency	53	332	332
Planned	199	1517	1592
FM/PDM	60	138	120
Contracted	10	96	80
	322	2083	2124

Payroll Data:

Straight Time Hours	1896
Overtime Hours	228
Total Hours	2124

Figure 8-16 Completed work analysis.

Performance Indices

A. Schedule Compliance

	This Week	Last Week
Total Hours Scheduled	1920	1880
Scheduled Hours Worked	1712	1626
Compliance (%)	89%	86%
B. Emergency Rate (%)	16%	20%
C. Overtime Rate (%)	12%	15%

D. Comments

  1. The main filter scraper arm has broken three times in the last 2 months. Material testing is being done to determine if metal fatigue is the cause.
  2. Emergency repackaging of the high-pressure pumps caused almost all of the schedule breaks. Until a solution is found for these pumps, a PM schedule is being set up.

Figure 8-17 Performance indices.

Graphical Presentation

When used in maintenance reports, graphical presentations tend to direct the focus of the reader. When combined with indices and raw data, the reader is more inclined to think of improvements or solutions to the problems presented. Graphical presentations convey complex data in a simple form. The goal in good graphics is to provide the viewer with the greatest number of ideas in the shortest time with the least ink in the smallest space.

Ideas conveyed when presenting maintenance data indicate that things are improving or things are not improving. Managers reviewing volumes of data tend to gloss over some information they may need to know. The time that managers spend on data interpretation can be better spent getting the proverbial 1000 words from a graphic than from reading a 1000-word report.

Summary

Professional credibility is achieved when you can prove the truth behind your statements with supporting data. Developing raw data into a useful index requires some talent, experimentation, and patience. Indices must be reviewed over time to see whether they provide insight into the operation of the maintenance department. An index that proves to be very useful at one kind of facility will be useless at another.

Whenever an index is used in a report, it is important to present the information truthfully. This usually requires that more than one index be presented to show a trend or to indicate an interrelation between indices.

A backlog index can be an indicator of the ability of the maintenance department to keep up with the new work that is generated every day. It can also be used to determine the proper size of the workforce. The backlog of maintenance work is the total estimated labor hours for all planned work. An index can be developed that gives one number to the backlog, expressed in weeks. The workable backlog index is calculated from a total of all the work that can be done immediately because parts and other materials are on-site. The total backlog is calculated from the work that is estimated and planned, as well as from parts and materials that are on-order or are available now. Backlog age is a measurement of the number of work orders in the backlog that can be completed within their priority period and the number that have missed the deadline. It is sometimes expressed as a percentage of the total number of work orders in the backlog that have aged beyond their priorities.

Schedule compliance is the value representing maintenance jobs actually completed according to the established schedule. This is usually expressed as a percentage of hours scheduled. Ideally, schedule compliance would be 100 percent. In the real world, where schedule breaks and emergencies occur, 90-percent schedule compliance can be an achievable goal.

Estimating accuracy is determined by dividing the actual hours of completed work by the estimated hours. Emergency work and other jobs that must be estimated on the fly (because of a perceived urgency) should not be included in this calculation. Estimating accuracy is often expressed as a percentage of the total actual work to total estimated work completed over a set period. Estimating accuracy should fall within the range of 85 percent to 115 percent.

Actual versus estimated comparisons can be used by a maintenance manager as one indicator of the success of the planning program. Some of the issues to examine are possible overstaffing, bad labor estimates, lack of parts, equipment still running, and poor work performance.

The easiest PM indicator is the percentage of PM work completed over a given period. If PM accounts for most of the completed work at a facility, emergency work is usually minimal. If little or no PM is performed, emergency work will be commonplace. Most industrial facilities have a running PM percentage between 20 percent and 40 percent.

The overtime percentage should be calculated and plotted on a weekly basis. A 3-month moving average may be helpful when it comes to identifying longer-term trends. Keeping overtime down to a minimum is vital in managing a budget. However, occasional overtime is unavoidable.

Productivity of the maintenance workforce is easier to control than it is to measure. Very few indicators exist that truly show the productivity of the workforce. Some common indices used are ratio of maintenance labor costs to maintenance material costs, maintenance cost per unit of production, and actual maintenance cost compared to budget costs

Although maintenance indices are not meant to relieve the maintenance manager from actual hands-on management of assigned areas of responsibility, these indices can point to areas on which attention must be focused. One index by itself will rarely identify an area of concern. Several indices should be evaluated to qualify a situation.

It's a good idea for the maintenance department to develop a weekly report of the activities. The report should consist of a balance sheet, performance indices, and graphical presentations.

Chapter 9 discusses how to improve productivity through the use of multiskill training.

Chapter 9

Using Multiskill Training

Nothing has affected industry more than the invention of the induction motor. Invented by Nicola Tesla at the end of the last century, the induction motor provided a very effective conversion of AC electrical power to rotating energy. Prior to its emergence, the use of water wheels and steam engines limited the size and complexity of factories. The induction motor allowed large manufacturing and processing centers to be built.

Along with this quantum leap in factory size, came an accompanying leap in complexity. Mechanical equipment grew larger and ran faster. Electrical distribution systems got much bigger to fill increasing energy needs. Further developments in electronics allowed more complex process-control instrumentation. This equipment allowed processes to be more under control at higher temperatures and pressures.

All these changes forced the specialization of skills. Many of the operators of equipment could no longer repair it. Master mechanic trades and utilities trades began to spring up in factories. This specialization continued, creating further separation into machinist, millwright, pipe fitting, welding, electrical, and instrument trades. More recently, an electronics technician trade was developed from the instrument and electrical trades.

During this burgeoning growth, the average industrial worker was educated through the elementary grades. The creation of specific *craft lines* seemed to be the only practical way to match complex machinery needs with the limited knowledge and training of those available to build and repair it. In the first half of this century, these craft distinctions became almost inviolate.

In today's age, craft lines do not need to remain so clearly drawn. The average industrial worker usually has a full high-school education, with many continuing on to further their knowledge with college-level and trade-enhancement courses. Many of these individuals even feel stifled by distinct craft lines.

Businesses today must be ready to compete globally. Improved transportation and communication has shrunk the world. The talk is now of global pressures and world market requirements. Multinational companies now blur the boundaries of industries. Single nations can no longer protect their industries, products, and workers from the downward pressure on prices and wages.

American industry distinctly feels this pressure. As former Third-World nations enter the global market with cheap labor, American products are often unable to compete. Rather than cut wages, move out of the country, or go out of business, most American companies have decided to reduce costs by *improving productivity*.

The food industry has historically built its business around *brand name familiarity*. Marketing strategies are built around the brand concept with the hope that increased share will be gained by imprinting the product name even deeper into the consumer's consciousness. But even brand loyalty has its bounds. Competitors are now lowering prices or selling through discount houses with pricing attractive enough to lure even the most faithful to possibly consider substituting their regular brand. Again, the only alternative left for this industry is to improve productivity so that the brand can remain competitive.

One opportunity being implemented as a real means of improving productivity is that of *multiskilling* efforts.

Understanding Multiskilling

Multiskilling is the process of training maintenance employees in specific skills that cross the traditional trade or craft lines, and then ensuring that the work is performed. The advantage of multiskilling is that particular jobs that historically required more than one craft (not necessarily more than one individual) are now performed by just one person.

A typical example is the change-out of a small motor. Traditionally, a change-out could require an electrician to disconnect the motor leads and a millwright or mechanic to disconnect the coupling, physically replace the motor, and perform the alignment. The electrician would then return to the job, reconnect the motor leads, and check (and possibly change) rotation. The mechanic or millwright would, at this point, be able to connect the coupling halves to complete the job.

In fact, no more than one individual should be required on this job at any time, but trade distinctions often require the close scheduling of appropriate crafts. If the loss of this motor created downtime, both individuals would remain at the job site, performing only their particular job functions as needed. In trade craft-dominated work environments, this situation may be even further complicated. The requirement for an operating engineer to physically remove and replace the motor may also exist.

In multiskilling, individuals would receive additional training beyond the normal skills required for their craft. The mechanic or millwright would be trained in the proper disconnecting and reconnecting of the motor leads, as well as how to change motor rotation.

The electrician, in turn, would be trained in coupling disassembly and reassembly, as well as alignment methods. After this training, either individual would be qualified to perform the entire job alone.

The multiskilling advantage to the company comes with the ease of scheduling work that, in the past, required two or more crafts or skill distinctions. The advantage to the worker is usually an incremental increase in pay for the additional skills learned and used.

Multiskilling Pitfalls

Many multiskilling efforts have not worked as desired. They usually fail for one or several of the following reasons:

- Unrealistic expectations
- Vague goals or commitments
- Inadequate definition of multiskilling
- Failure to implement

Unrealistic Expectations

When company management and the hourly workforce begin to discuss the possibilities of multiskilling, both parties usually enter into this dialogue with the following expectations:

- The company, feeling hampered by the requirement to send out two people on many one-person jobs, or frustrated over the effectiveness of tight scheduling of multiskilled work, envisions the development of a race of super mechanics—individuals who can do anything and everything.
- The hourly workforce, after hearing the company's expectations, dreams of a new day in which their paychecks will be bursting at the seams from a new pay scale—the end result of having been trained in all the craft areas instead of just one.

Neither expectation is realistic. Although there may be one or two exceptional individuals who can learn and retain many skills, most will not attain this level. In addition, the ability to retain a skill is dependent on being able to use the skill on a regular basis. Ensuring that all workers trained with many skills will regularly exercise those skills could be more of a scheduling nightmare than the original problem.

For the hourly worker, a multiple-wage scale is unrealistic for economic reasons. Though some increase in pay scale may be realized because of the higher productivity benefit of multiskilling to the company, there will be a ceiling to that increase after which the cost of extra pay will cancel any productivity savings.

Vague Goals or Commitments

Many multiskilling efforts begin without clear goals or understandings by the company of what it expects to gain. In turn, there is no commitment exacted from the workforce of what will be expected once the training has been accomplished. The end result is that after the training has been finished, the manner in which maintenance work is scheduled and performed remains the same as it was before the training effort was begun.

Inadequate Definition of Multiskilling

Even if there has been an attempt to define the goals of multiskilling and the workforce is committed to the end result, the training must adequately fit the goals. Inadequate definition of the training effort itself can leave the goal unclear in everyone's minds as to when and how the effort will be accomplished. Again, the end result is that the scheduling and performance of maintenance remains relatively unchanged.

Failure to Implement

Strange as it may seem, after negotiating an agreement for multiskilling, many companies have failed to quickly follow through with the plan or failed to communicate the goals, definition, and work changes in the multiskill agreement. As a result, the new level of productivity that was initially desired is not achieved and the advertised benefits are never realized.

A process of multiskilling must be developed and begun prior to the negotiations for the program.

Successful Multiskilling Programs

It may be evident that a successful multiskilling effort must be well defined from the very beginning. The company and its employees must be of one mind with respect to the following:

- What training will be required for the effort?
- What skills will be involved?
- What work will be covered?
- How and when will the work be executed?
- What specific benefits will be expected?

Answering these questions begins the process of multiskill development. Many of the requirements of multiskilling can be determined by identifying the friction areas.

Identifying Friction Areas and Multiskill Opportunities

In plants or facilities in which strong craft line distinctions exist, it is imperative to identify the skills to be included in the program. The most productive areas to be considered are those that involve jobs where two or more crafts are required, but only one or two individuals are required to do the work. These are best identified as *friction areas*—jobs that are causing friction in the productive deployment of personnel. These friction areas can be identified through several means.

- *Review completed work history records*—Work histories will often indicate jobs that require more skills than people normally have.
- *Brainstorming sessions*—Include first-line supervision to identify potential multiskilling areas.
- *Structured group interviews*—This group-interaction tool allows for positive input and prioritization of that input for a group of people not normally involved in brainstorming sessions.
- *Multiskill survey*—Using a questionnaire completed by all management and affected hourly personnel, a survey is designed to poll individuals on their willingness to cross-train, ideas for multiskill opportunities, and perceived problems with a multiskill implementation. Figure 9-1 shows a sample survey. This survey should be modified to fit the needs of each location.

Potential Friction Areas

Although the friction areas in which multiskilling will be considered can vary greatly from location to location, the following areas are commonly considered when multiskilling is being looked at as a productivity-improvement area:

- Jobs combining electrical and mechanical skills (motor change-outs, some instrument modifications)
- Jobs requiring electrical/mechanical and simple welding skills (installing conduit/pipe supports and running conduit/pipe)
- Pipefitting work (where pipefitter and welder are separate craft skills)
- Minor insulation removal and replacement (insulation work associated with piping or leak repair)

MULTISKILL SURVEY

Please complete the following questionnaire with respect to a proposed multiskilling effort.

Employee Name: _____ Date: ___/___/___

Check the appropriate box:

	Strongly Agree	Agree	Don't Know	Disagree	Strongly Disagree
1. Multiskilling would benefit our maintenance effort	☐	☐	☐	☐	☐
2. Employees might lose their job due to multiskilling	☐	☐	☐	☐	☐
3. Multiskilling will create unsafe situations	☐	☐	☐	☐	☐
4. The process of multiskilling is understood by most plant employees	☐	☐	☐	☐	☐
5. Multiskilling is not achievable here	☐	☐	☐	☐	☐
6. Multiskill training alone would enhance the skills of current employees	☐	☐	☐	☐	☐
7. Most craft employees do not want to perform other jobs	☐	☐	☐	☐	☐
8. The company is capable of managing a multiskill effort to the benefit of all employees	☐	☐	☐	☐	☐

What specific maintenance skills would make you more productive in your job? How often in the past year could you have used each skill?

Maintenance Skill	Times Used in Past Year

Figure 9-1   Multiskill survey.

- Forklift operation (assuming that forklift driver is a bid position)
- Minor rigging operations (where cherry-picker operation is a bid position)
- Minor machining operations (turning down, reaming, broaching)
- Oxyacetylene operations (cutting, trimming, heating)
- Machine lubrication (refilling after rework)

Identifying Potential Gains

Once the possible training areas have been identified, the company can determine the potential productivity and financial savings to be achieved from the multiskilling effort. The financial savings can be shared with craft employees through negotiated wage increases. This effort takes the following form (in order):

- Interviews are conducted with supervisors to identify friction areas.
- Completed work history is reviewed for friction areas and these jobs are tabulated.
- A study is conducted as to how these jobs could be performed under a multiskill arrangement.
- Estimates are made for the hours that could have been saved through a multiskill effort on specific jobs. A calculation of labor cost savings is performed.
- Tabulation is made of any productivity improvements resulting from reduced clock hours of downtime. The cost of lost production is calculated.

Possible wage increases can now be determined by examining all information accumulated and negotiations with workforce representatives can begin.

Defining the Training

Defining the multiskill training is the most important step in the effort. The training must equip workers with the specific skills they will need to safely perform the duties formerly accomplished by another craft. As with all maintenance training, the curriculum of the multiskill training effort must include the following elements:

- Coverage of technical aspects of the training topic.
- Coverage of safety aspects of the training topic.
- Hands-on performance of the training topic, with appropriate assessment and correction.
- Performance acceptance, where trained employees are given the chance to demonstrate new skills.

In many cases, this training will take the form of *spot training,* designed to equip an employee with a critical skill (that is, alignment, motor connection, welding, and cutting).

In addition to multiskill training, a *training progression program* can be developed. It is unlikely that skilled individuals hired from the outside would have all the combined skills required in a multi-skill environment. Also, individuals promoted to maintenance from operations or the labor pool may not be equipped with many of the skills required. A training progression program designed to bring an employee to the full multiskill level can ensure continuation of the effort in years to come.

Negotiating the Multiskilling Program

At some point in the development of the multiskilling effort, the company and workers will have to sit down and negotiate the benefits to both sides. This process is made much easier when there has been a clear definition of the specific areas in which multiskilling will occur. The process itself varies depending on the working relationship between the company and workers, but usually covers the following areas:

- General definition of areas where multiskill training will be performed.
- Definition of the *demonstration of performance* when training has been completed.
- Incremental pay increases, which will accompany the training effort.
- Grandfathering, if necessary, of any workers currently in the maintenance workforce who simply cannot learn another craft skill. This number must be kept to a minimum to ensure that the multiskill approach works.

Implementing the Skills

Although it would almost seem intuitive, it is important to identify exactly when and how the multiskilling skills will be incorporated into the actual scheduling and work performance. Failure to do this has often short-circuited the entire effort, with the company incurring the cost but never realizing the benefit.

Multiskilling—A Win-Win Effort

Many companies and the maintenance personnel who work for them are in a potential win-win situation. Multiskilling offers real productivity gains for any company, and the extra skills and job enrichment can mean a boost in the paycheck and the self-esteem for those performing the work. Successful programs, however, are cognizant of the potential pitfalls and are designed to avoid those problems. Development of a quality multiskilling effort provides tangible benefits for all involved, and can be an area for major productivity improvement for many manufacturers today. Figure 9-2 illustrates the approach to multiskilling as it has been explained in this chapter.

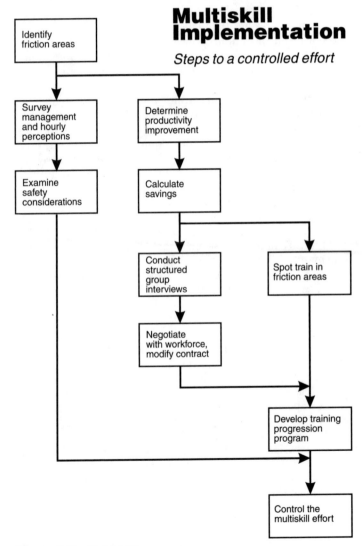

Figure 9-2   Multiskilling process.

Summary

Businesses today must be ready to compete globally. Improved transportation and communication has shrunk the world. Single nations can no longer protect their industries, products, and workers from the downward pressure on prices and wages. American industry distinctly feels this pressure. As former Third-World nations enter the global market with cheap labor, American products are often unable to compete. Rather than cut wages, move out of the country, or go out of business, most American companies have decided to reduce costs by improving productivity. One opportunity being implemented as a real means of improving productivity is that of multiskilling efforts.

Multiskilling is the process of training maintenance employees in specific skills that cross the traditional trade or craft lines, and then ensuring that the work is performed. The advantage of multiskilling is that particular jobs that historically required more than one craft (not necessarily more than one individual) are now performed by just one person. In multiskilling, individuals receive additional training beyond the normal skills required for their craft.

The multiskilling advantage to the company comes with the ease of scheduling work that, in the past, required two or more crafts or skill distinctions. The advantage to the worker is usually an incremental increase in pay for the additional skills learned and used. However, many multiskilling efforts have not worked as desired. They usually fail because of unrealistic expectations, vague goals or commitments, inadequate definition of multiskilling, or failure to implement.

In plants or facilities in which strong craft line distinctions exist, it is imperative to identify the skills to be included in the program. The most productive areas to be considered are those that involve jobs where two or more crafts are required, but only one or two individuals are required to do the work. These are best identified as friction areas—jobs that are causing friction in the productive deployment of personnel.

At some point in the development of the multiskilling effort, the company and workers will have to sit down and negotiate the benefits to both sides. This process is made much easier when there has been a clear definition of the specific areas in which multiskilling will occur. The process itself varies depending on the working relationship between the company and workers.

Appendix A

Identifying Electrical Work to Be Performed During a Shutdown

When shutdown or turnaround is planned for industrial facilities, a plan for electrical maintenance that requires an electrical outage is usually not included. This is unfortunate, because electrical outages provide the planner with a unique opportunity to define future required work. This appendix includes a partial list of work performed before and during an electrical outage.

Table A-1 describes tasks that should be performed before an outage, and Table A-2 describes tasks to be performed during an outage.

Table A-1 Before the Outage

Step	Task	Description
1.	Conduct an infrared scan.	A standard three-phase power system will usually have equal current flowing in all three phases. As a result, all heat given off by the cables, contacts, and connections should be equal in adjacent phases. If a connection is loose, corroded, or dirty, it will heat up more than the other phases. The infrared scanner provides a method for finding these hot spots, which, if left uncorrected, could result in catastrophic failure. If a possible problem is identified with the scan, an infrared thermometer is used to determine the severity. A 5°C increase above normal operation is considered to be a point of concern.
2.	Test the transformer oil.	Transformer oil samples can be taken from most oil-filled transformers while they are running. Two types of tests should be performed on the oil: basic screening and dissolved gas analysis Basic screening tests determine the serviceability of the oil as a cooling medium and as an insulator. Dielectric, interfacial tension, and acidity are some of the tests performed.

(continued)

Step	Task	Description
		Dissolved gas analysis (or gas-in-oil) can reveal something about the condition of the transformer windings, internal connections, and iron core. For example, the presence of combustible gases (such as carbon monoxide) indicates destruction of the winding-phase separators. Acetylene, in even very small quantities, can indicate a high-energy arc in the transformer. When high combustibles are found, retests are usually performed to see if the gases are increasing.

Table A-2   During the Outage

Step	Task	Description
1.	Test overcurrent on molded case breakers, thermal overload protective devices, and other protective devices.	Overcurrent devices have published *time vs. current characteristics*. Trip tests are done to determine if a protective device is working within its performance rating.
2.	Test megohm of rotating machines, transformers, insulator bushings, and power cables.	The megohmmeter is an instrument designed specifically to measure insulation resistance directly in megohms (one megohm equals 1,000,000 ohms). Insulation resistance can be measured without damaging the insulation and furnishes a highly useful guide for determining the general condition of insulation. Megohmmeter testing may be used to test the insulation resistance between conductors of separate circuits or between the conductors and ground. The resistance measured in *good insulation* will begin very low and then begins to level off after about 10 to 15 minutes. If the insulation is *wet or dirty*, the resistance will level off quickly. There are two insulation values that can be used to ascertain the quality of the insulation: the spot reading and the polarization index.

(continued)

Table A-2   (*continued*)

Step	Task	Description
		The *spot reading* is the megohm reading after one minute. IEEE Standard 43 establishes a limit for this spot reading as the Kvolt value of the device under test plus 1 (in megohms). In other words, for a 460-volt motor, the insulation resistance limit would be 1.46 megohms. This value may be acceptable as a *troubleshooting limit*, but would prove to be unacceptable if extended service is to be ensured. Insulation measuring 1.46 megohms could last 6 months to 1 year, or could fail in a day.
		A better limit is 100 megohms. New insulation has a megohm reading close to 2000 megohms. When this value degrades to 100 megohms, there is still time to repair the damage. If it is allowed to degrade further, to 1.46 megohms, it would be close to failure.
		Another test measures the dielectric absorption. A dielectric absorption test is an extension of insulation resistance testing for longer than the conventional 1-minute period. To perform this test, a motor-driven or battery-operated megohmmeter is required. In this test, the insulation resistance is recorded at 1 minute and 10 minutes when readings plateau. Clean and dry insulation in good condition will steadily increase in resistance value over the 10-minute period. Dirty or moist insulation will plateau quickly and at a relatively low value of resistance.
		A number called the polarization index provides an easy way to evaluate the results. The *polarization index* is the ratio of the 10-minute reading to the 1-minute reading, calculated as shown here:
		Polarization index (PI) = $$\frac{R10 \text{ (Resistance at 10 minutes)}}{R1 \text{ (Resistance at 1 minute)}}$$
		If the insulation resistance reading *more than doubles* itself from 1 minute to 10 minutes, then the insulation is considered to be in good condition.

(continued)

Table A-2 (*continued*)

Step	Task	Description
		If a motor has a poor polarization index, it usually does not require a rewind. In most cases the motor can be dried out and varnished, resulting in a much-improved megohm spot reading and polarization index.
3.	Test DC high potential.	The dielectric strength of an insulation system determines the level of voltage a piece of equipment can withstand without arcing over to another phase or to ground. It also determines how much overpotential (such as a voltage surge) the insulation can withstand. Dielectric strength testing is accomplished by placing a calculated voltage on conductors to test their insulation. This may be either an AC voltage using an AC high-potential (hipot) tester or a DC voltage using a DC hipot tester. DC high potential testing is preferred for maintenance use, because, if performed properly, it can be nondestructive. AC high-potential testing is a more sensitive procedure and the possibility of a poor device or cable failing under the test is much greater.
4.	Mechanically operate and inspect large breakers.	Large circuit breakers should always be removed and inspected during electrical outages.
5.	Clean switch gear and other cubicles.	Switch gear and all other cubicles should be thoroughly cleaned during an outage. Debris left behind might eventually lead to a major problem.

Electrical Shutdown Checklist

The remainder of this appendix provides a checklist of information to be collected for an electrical shutdown. The information is broken down into two major milestones:

- Months and weeks prior to the shutdown
- During the shutdown

Months and Weeks Prior to the Shutdown

The following list shows all work that should be investigated (in the order of priority) before any plant shutdown is to occur:

- Poles and cable support structures
- M/V breakers + fuse disconnects
- L/V switchgear (metal clad)
- Transformer checks
- Infrared scan
- Motor starters and MCCs
- General shutdown checklist

The following sections show information that should be gathered at each stage of the shutdown.

Poles and Cable Support Structures

General

Walking the system periodically is the key to inspections (use binoculars where necessary). Aerial cable installations should be inspected for mechanical damage caused by vibration, deteriorating supports, or suspension systems.

Detail

Note all discrepancies for correction during a future shutdown.

1. Infrared scan to be performed on all cables and connections. See the section titled *Infrared Scan*, for details.
2. Inspect static lines and guide wires for corrosion, abrasion, tension, and tight connections to supports.
3. Inspect poles and cross-beams for structural integrity. Identify broken insulators.
4. Potheads should be inspected for oil or compound leaks.

M/V Breakers + Fuse Disconnects

1. Inspect enclosure for dust, dirt, rodents, reptiles, corrosion, and tracking.
2. Record the number of operations (if counter is present) and check indicating lights.
3. Reset any tripped relay flags. Investigate reason for flag and determine if a long-term corrective action is necessary.
4. Operate space heaters. Use amp probe to determine if they are operating properly.

5. Inspect batteries for proper water level and low specific gravity. Clean as necessary.

6. Contract for overcurrent and other relay testing (annually).

L/V Switchgear (Metal-Clad)
Detail

1. Test voltage with DVM, phase-to-phase, and phase-to-neutral to identify any grounds. Record all readings. Verify ground lights are operating. Voltages should be within 1 percent of each other from phase to phase. All readings should be + or − 2 percent of transformer nameplate rating (that is, 480 volts + or − 9.6 volts). Investigate any irregularities.

Actual Voltages:

L/V Switchgear Name: _____

A to B _____ volts	A to G _____ volts
B to C _____ volts	B to G _____ volts
C to A _____ volts	C to G _____ volts

2. Infrared scan for loose or corroded connections. See the section titled *Infrared Scan*.

3. Contract for overcurrent tests to be performed on breakers (at least every 3 years). Be sure spare breakers are available before the shutdown in case one needs to be replaced.

Transformer Checks
Detail

1. Record primary and secondary volts and amps. Verify that current is below transformer ratings on all phases. If current exceeds self-cooled or partially force-cooled levels, be sure proper cooling equipment (fans and/or pumps) is operating to dissipate heat.

Voltages should be measured either with existing meters or a DVM. Phase-to-phase or phase-to-ground readings are both acceptable. Readings must be between + or − 5 percent of rated voltage (tap setting value). Abnormal voltages indicate possible supply voltage changes and the utility should be contacted. Three-phase voltage unbalance is also a possible problem and abnormal changes should be investigated. Unbalance of 1 percent or greater use is severely detrimental to electric motors.

Actual Voltages:

Transformer Name: _____

A to B _____ volts	A to G _____ volts
B to C _____ volts	B to G _____ volts
C to A _____ volts	C to G _____ volts

2. Record hot-spot and top-oil temperatures. (Hot-spot thermometers are usually found on large transformers only.) At peak loads, the transformer hot-spot temperature should be on the high end of its normal operating range. Hot-spot temperature at low loads will be lower than at peak. Top-oil temperatures may differ only slightly from inspection to inspection (because of the forced cooling system). Any major change in top-oil temperature is an indication of either cooling fan or pump failure, improper fan and pump control settings, or major failure of the transformer oil.

Temperatures at 40°C Ambient, Upper Limit

	Gauge	Drag Needle* (High-point)
Hot spot	65°C	No greater than 105°C
Top oil	55°C	No greater than 95°C

(on most transformers)

Actual Transformer Gauges

Transformer Name: _____

	Gauge	Drag
Hot spot	_____°C	_____°C
Top oil	_____°C	_____°C

3. Record oil level. Oil levels will be slightly lower at low load periods and on cold days. Be sure to mark this lower limit on the level gauge as a guide to any loss of oil in the transformer.

   Sight glass–type gauges should be full to a level just below the transformer tank top. Magnetic type gauges will have a LO range marked on them.

Actual Oil Level

Transformer Name: _____

Level Gauge _____ (HI, LO, NORM, MID, and so on)

4. Nitrogen blanket system should be checked for maximum and minimum pressures (for maintained blanket only).

---

* The drag needle is the needle on the gauge that is pushed to the highest transformer temperature by the indicating needle. This needle should always be reset to the indicating needle position after inspection.

Limits

> **Blanket Pressure**
> +8 psi gauge maximum to +1 psi gauge minimum
> **Nitrogen Tank Pressure**
> 150 psi minimum
> Any pressure outside the limit should be investigated immediately.

5. Inspect for physical damage. Note any dents, scratches, loss of paint, or corrosion on the transformer and associated equipment. Inspect transformer mountings for sagging or deterioration.

6. Note any leaks from transformer tank, fittings, cooling tubes, and bushings. Some leaking of transformer oil is common during initial start-up only. Any leaking component should be immediately cleaned and the condition should be monitored. If leaking persists, the transformer must be de-energized and the source of the leak corrected.

7. Verify proper auxiliary device operation. Operate (on manual) all pumps and fans. Note proper operation of flow indicators (for pumps). Identify any excessive vibrating device. Shut down all fans, inspect and clean dirty fan blades.

CAUTION

Be sure temperature limits are not exceeded while fans are down.

8. Take vibration tests (in ips velocity) on transformer and associated equipment. The transformer tank readings should be consistent with previous readings while noting transformer loading. (Mark a spot on the transformer for consistent readings.) A 20-percent increase above initial (new) readings of vibration are an indication of mechanical looseness in the transformer or core failure.

   The motors, pumps, and fans *must not* have vibration levels of 0.3 ips or greater. These readings should be taken at the equipment bearing housing.

9. Oil and operate any and all locks on the transformer, including manual tap changer locks. Do not operate tap changer.

10. Inspect and tighten external ground connections.

11. Perform infrared scan of transformer. (See the section titled *Infrared Scan*.) Identify any loose or corroded connections, overheated bushings, or blocked cooling fans. Over-temperature

bushings or connections are detected by uneven heating on the three-phase transformer connections. Blocked cooling fins are identified by a cooler (or dark infrared) fin in a group of fins.

12. Perform basic oil screening and dissolved gas analysis on all plant transformers. Contract this service from a reputable transformer oil tesing company.

13. Contract repair service if required.

Infrared Scan

General

An infrared study is performed for the purpose of detecting loose or corroded connections on electrical equipment. Electrical current flowing through copper or aluminum conductors generates a measurable amount of heat. This heat is mainly caused by the resistance to current flow that exists in all metal conductors. The heat generated, or power lost, as it relates to resistance (R) and current flow (I) is described as follows:

$$\text{Power Lost} = I^2 \times R$$

If a loose or corroded connection should occur, the resistance to current flow increases greatly. So does the generated heat. This heat may be enough to distort and melt the connection, resulting in an electrical outage. This work is usually performed by an outside service because of the expense of the equipment required. An infrared imager with some sort of recording (such as a photo) and a temperature detector are required for this survey.

Detail

1. Conduct an inspection and document all equipment and locations where an infrared scan can be performed. Items usually considered are:

- Aerial cable
- All cable connections
- Bus work and connections
- L/V substations
- Motor control centers
- Motor starters and C/B connections
- Transformer cooling fins

2. Install hinges or handles on equipment that may be difficult to access while hot. Often the installations of handles on large switchgear enclosures make access quicker and safer during infrared inspections.

3. Schedule the infrared scan for a period in which load in the plant is near maximum. This will allow the worst problems to better exhibit themselves.

4. Perform infrared scan. Abnormal heat levels can more easily be detected in three-phase power systems. Most three-phase power connections will have an equal amount of electrical current flowing through each phase (hence, an equal amount of heating on all connections). If a loose or dirty connection should occur, the same current would flow through all three phases, but more heat would be generated at the loose connection (because of increased resistance, or $-R$). Eventually, unbalanced voltages or even single phasing may occur, causing failure not only at the faulty connection, but also to all other equipment connected to it.

The infrared viewer can detect any unbalanced heating in a three-phase load and, using the measured temperature, the severity of the problem can be deduced. All three-phase electrical connections and cables should be at a temperature within 5°C of each other.

The following items must be recorded on problems found through the infrared scan:

- Ambient temperature of the equipment in the area
- Temperature of all three phases (if a three-phase system)
- Current through the connections or cable
- A drawing or sketch of the equipment with temperatures identified
- Date when problem was found
- Recommended latest repair date
- Photos of all problems

Motor Starters and MCCs
Detail

1. Inspect all starter cubicles in the area to be shut down for the following:
   - Dirt, dust, rodents, spider webs, and so on.
   - Corrosion on starter components or cubicle.
   - Noisy coils or contacts.
   - Burnt or distorted current carriers.

2. Perform an infrared scan on all operating starters. (See the section titled *Infrared Scan.*)

3. Make a list of all associated motors, and record all nameplate data.

4. Make a list of all parts and equipment that will be required for the test.

5. Inspect all junction boxes and cable trays.

General Shutdown Checklist (Suggested List)

1. Where will temporary power be needed?

2. Will emergency generators or temporary generators be needed?

3. What special tools/equipment will be needed?
   - Compressors
   - Temporary generators
   - Scaffolding
   - Lifts
   - Cranes
   - Special instruments

4. What contractors or consultants need to be called in?

5. What spare parts or supplies need to be purchased (or check stores)?

6. List all work orders to be performed during this shutdown.

7. List all purhase order numbers to be issued for the shutdown.

8. Make a plan. What needs to be done first? What jobs must you coordinate with other crafts or production? In what sequence should the work be done?

9. Make sure everything (instruments, equipment, and so on) is in working order prior to shutdown.

During the Shutdown

With equipment is ready for maintenance or de-energized, be sure to gather data for the next shutdown and identify parts and materials needed.

L/V Breakers (Metal-Clad)

General

L/V Breakers should be removed from service and tested at the designated interval or after the breaker has opened to interrupt a damaging fault. Stored energy should be discharged prior to any maintenance.

Detail
All performed annually.

1. Remove arc chutes. Inspect, adjust, and clean as necessary for broken parts, missing arc splitters, metal splatter, and burning on interior surfaces. Snuffer screen must be clean. Replace arc chutes with new ones, if required.

2. Inspect main contacts for pitting, spring pressure, overheating signs, alignment, over-travel, or wipe, evidenced by slight impressions on the contact surfaces. Clean or replace as necessary. Silver contacts will show more discoloration and indentation than copper. This is usually acceptable. Never file contacts. Nothing more abrasive than crocus cloth or Scotch Brite should be used.

3. Inspect arcing contacts for alignment, over-travel, or wipe (evidenced by slight impressions on the contact surfaces), and for arc erosion. Arcing contacts are the last to open during a fault. Verify this by slowly manipulating the operating mechanism. Replace as necessary.

NOTE

Contacts should be replaced in threes.

4. Disconnect finger clusters. Inspect for proper adjustment and spring pressure.

5. Check structure or frame for proper alignment and loose or broken parts.

6. Inspect all insulating materials for cracks, breaks, or signs of overheating.

7. Oil all parts such as bearings, racking screws, and sliding screws. Do not apply lubricant to contacts.

NOTE

The following tests should be performed by competent personnel. Test equipment required to achieve reasonable test currents may dictate the use of outside services. (Greater than 30,000 amps of continuous test current is sometimes required.)

8. Contact resistance test (breaker closed):
   Use a Ductor, Wheatstone Bridge, or other micro-ohmmeter to measure contact resistance on all three phases (from input to output). Resistance shall not be more than 2000 micro-ohms,

and values should not deviate from pole to pole, or from similar breakers, by more than 50 percent. Record all readings.

9. Insulation resistance test (breaker open):

    Use a 1500-volt megohmmeter to test each phase (on the load and line side) to ground (case), phase to phase, and phase line to phase load on each phase. Readings must not be below 1000 megohms. All readings should be recorded at 60 seconds.

10. Overcurrent trip testing (electromechanical and solid-state type).

    Using a high-current test device, verify trip curves by testing electromechanical trip devices at 150 percent and 300 percent of rated current, and verify instantaneous trip setting. Test current to be injected at the contact fingers with proper size stab. Verify trip curves by testing solid-state trip devices. All existing trip settings should be verified.

NOTE

These tests, inspections, and adjustments should be performed in conjunction with L/V breaker tests. Verify that no parts of the power or control circuitry are energized by backfeed from other power or control sources.

11. Completely clean with vacuum and lightly dampened rags.

12. Clean and repaint enclosure as necessary. Clean all vents. Oil hinges.

13. Check bolts, bus connections, and splices for required tightness. When in doubt, use the standard torque table.

14. Clean all insulators, insulating material, and inspect for cracks.

15. Clean and inspect breaker stabs for pitting and effects of over-heating.

16. Inspect cables and connections for evidence of over-heating or frayed insulation.

17. Clean, align (as necessary), and oil racking mechanisms.

18. Test all meters for accuracy. Repair or replace as necessary.

19. Verify all grounding is intact.

20. Test conductivity of all aluminum cable and bus connections. Use a Ductor, Wheatstone Bridge, or other micro-ohmmeter to measure connection resistance. Resistance must be below

2000 micro-ohms. Use Belleville washers when bolting aluminum cable lugs to equipment.

21. Use a 500-volt megohmmeter to test each phase on the main bus from phase to ground and from phase to phase. Measure resistance on all incoming and outgoing power cables. All readings should be taken for 60 seconds. All readings must be above 1000 megohms.

22. Reinstall breakers and verify that disconnect finger clusters are making proper contact with stabs. Do not lubricate contacts. Measure contact resistance when proper contact is in doubt.

Low- and Medium-Voltage Motor Starter Test and Maintenance
(See data collection sheet.)

Detail

For low-voltage starters only:

NOTE
Be sure the motor is down and power to the breaker is off.

1. Clean and vacuum all cubicles. Oil all hinges. Do not lube starter components.

2. Disconnect the line and load leads from the breaker and perform the following megohm tests. All readings must be above 500 megohms.

Line side:
  Phase to ground, A phase, with B and C shorted to ground
  Phase to ground, B phase, with A and C shorted to ground
  Phase to ground, C phase, with A and B shorted to ground

Load side:
  Phase to ground, A phase, with B and C shorted to ground
  Phase to ground, B phase, with A and C shorted to ground
  Phase to ground, C phase, with A and B shorted to ground

Other:
  A line phase to A load phase
  B line phase to B load phase
  C line phase to C load phase

  All megohm tests should be performed for 60 seconds unless megohm reading is greater than 2000 megohms, in which case motor is good.

3. Remove the breaker and have it tested (by vendor).

NOTE

The tests in (4) and (5) are performed at 3 years. These tests are best conducted by an outside service. (Special test equipment is required to provide currents in excess of 30,000 amps.)

4. Using a Wheatstone Bridge, Ductor, or other low-ohm device, measure contact resistance of all breakers. Resistance shall not be higher than 2000 micro-ohms.

5. Overcurrent test all molded case circuit breakers and compare results with time–current characteristics. Emphasis must be put on instantaneous trip operation. If breaker does not trip on first test, raise test current above trip setting by 50 percent (but not to exceed breaker interrupting rating) and test again until unit trips. Test all three phases, recording as-found and as-left readings. Replace any breaker that fails to trip or trips earlier than design characteristics.

Detail
For low- and medium-voltage starters:

6. Meg out the starter as described for the breaker in step (2). All readings should be above 500 megohms (taken for 60 seconds unless it is above 2000 megohms). If the starter megs poorly, either remove and clean it, or replace it.

7. Inspect contacts if accessible. Clean or replace as necessary. Never file or use sandpaper on contacts.

8. Check arc hoods for cracks, breaks, or deep erosion.

9. Change cracked or embrittled coils.

10. Reinstall all starter cubicle parts and tighten all connections.

11. Paint cubicle if necessary.

Transformers (Oil Filled)

General All tests, repairs, and inspections in this section must be made after the transformer is completely discharged. All instrument transformers must be disabled either by removing fuses or shorting the secondary and primary windings.

Before work is performed in the transformer, all tools should be counted and listed. The list should be checked after the end of work to make sure no tool has been left in the transformer. All tools used in the transformer should be tied to a long string connected outside the transformer. Any person entering the tank should remove any loose items (change, keys, and so on) from pockets prior to entering. A transformer tank is just that—a tank—and a standard Tank Entry Permit must be completed before anyone is allowed to enter.

Some tests, repairs, and inspections may not apply to all transformers.

Detail
Annually

1. Perform required maintenance on the transformer as determined by inspections.

2. Contract oil servicing, as required.

3. Inspect pressure-relief diaphragm for cracks or holes and verify proper operation. (possible cause of pressure in sealed-type transformers remaining at zero).

4. Clean bushings and lightning arrestors and inspect surfaces. Check level and refill as necessary all oil-filled bushings. (Westinghouse type 0 bushings are oil filled.) Check for leaks, hardened bushing gaskets, and corroded or broken ground connections.

5. Inspect tap changer, following manufacturer's instructions and contact replacement frequency requirements.

6. Paint tank as required. Wire-brush rust spots and prime paint. *Never paint fan blades.*

7. Make undercover inspection. (Positive pressure should be applied to prevent moisture entrance.) Note moisture or rust under the man-way covers at the top of the tank. Also, note oil sludge deposits, loose bracing, or loose connections. Take corrective action as necessary. (Moisture or rust on the cover may be an indication of water in the oil. Oil screening test should be reviewed.)

8. Check all ground system connections, inside and outside the transformer, for corrosion. Correct as necessary.

9. Check *all* drain valves for proper operation. Check, tighten, or replace gauges, valves, and fittings.

10. Check tightness of all electrical connections on transformer.

11. Test insulation resistance (do not perform this test if the transformer is emptied of oil). Using 5000-volt megohmmeter, conduct time-resistance tests for insulation resistance. Prior to test, note insulation temperature (with either infrared thermometer or resistance method—see Item 12). All spot readings should be corrected to 40°C base.

12. Test should be performed on all windings to ground and between windings. Guard circuits must be used to drain unwanted leakage.

The following describes all possible combinations on a three-winding transformer that should be checked:

- High to low, tertiary, and ground at high voltage
- High to ground at high voltage
- Low to ground at low voltage, then high if it passes
- High to tertiary at high voltage
- Low to tertiary at low voltage, then high if it passes

Permanently connected windings should be considered one winding. All tests are to be conducted for a minimum of 10 minutes, noting insulation resistance at 30-second intervals.

Limits

$$PI = \frac{R \ 10 \ minutes}{R \ 60 \ seconds} = \text{should be greater than 2.0}$$

Resistance at 60 seconds (corrected to 40°C) should be equal to or greater than 500 megohms.

Correcting Insulation Resistance for Temperature

$$Rc = Rm \times KI$$

where the following is true:

Rc = Corrected insulation resistance
Rm = Resistance measured after 60 seconds
KI = Correction factor from the following table

°C (winding temperature)	KI	°C (winding temperature)	KI
0	0.06	45	1.41
5	0.09	50	2.00
10	0.13	55	2.82
15	0.18	60	4.00
20	0.25	65	5.64
25	0.35	70	8.00
30	0.50	75	11.28
35	0.71	80	16.00
40	1.00		

13. Low-resistance winding test.
    Using a Ductor, Wheatstone Bridge, or other low-ohm (micro-ohm) meter, measure and record transformer winding resistance. All high-voltage windings should read the same, as should all the low-voltage windings. All high-side winding

resistance should be within + 5 percent, as should all low-side windings. If the transformer has cooled to ambient air temperature at the time of the test, record the temperature indicated. Correct all resistance readings to 40°C by using published tables for copper or aluminum windings.

$Rs = Kw \times Rm$

where the following is true:

Rs = Resistance at desired temperature (40°C)

Rm = Measured resistance

Tm = Temperature at which resistance was measured

14. Turns ratio test:

This test is best performed by a qualified electrical testing contractor. The results of this test should verify the turns ratio of all transformer windings (that is, primary voltage divided by secondary voltage). Also, all transformer winding ratios should be within + 0.5 percent of calculated ratio.

Test results can be used in conjunction with winding resistance measurements to diagnose transformer problems. A positive turns ratio test with a negative winding resistance test indicates loose connections in the transformer, possibly in the tap changes.

15. Insulation power factor test (optional for transformers less than 2000 KVA).

This test is best performed by a qualified electrical testing contractor. The results of this test should be below limits established by Doble Engineering, which are:

Below 0.5 percent for a new transformer

Below 1 percent for a transformer that has been in service for 10 years or more.

16. Exciting current test (optional for transformers less than 2000 KVA).

This test is best performed by qualified electrical testing contractors. The pattern of test data should be looked for here. Two similar current readings for the outside phases and a lower current reading for the center phase of the transformer are indicative of good readings. Shifting of the transformer core laminations will result in a major change from this pattern or from previous readings.

Glossary

acceleration—The change in velocity with respect to time.

acoustic emission—The method to detect early stress cracks in vessels, reactors, storage tanks, and pipeline transmission systems.

activity—A task, job, or work order required in the completion of a project. Commonly used in critical path method (CPM) scheduling, identified with a line and an arrow with a short description or abbreviation of the task. An activity can also be represented by a box in a critical path method (CPM) network. See also *critical path method.*

addendum—A document explaining a change or correction to a contract.

aggregate—Small stones or gravel used in construction of mortar or concrete. An aggregate of stone is also used on built-up roofs to dissipate heat.

allocation scheduling method—A technique that refocuses the workforce on preventive maintenance (PM) and predictive maintenance (PDM) work, while providing a facility-wide priority system to complete the other important jobs in the facility.

American National Standards Organization—An organization that publishes references defining the methods, classification, and testing of materials and standard languages used in science and industry. See also *American National Standards Institute.*

American National Standards Institute (ANSI)—An organization that develops trade and communications standards.

area maintenance approach—A technique in which a separate (and often autonomous) maintenance workforce is assigned to each operating area (a processing segment or type of operation).

arrow diagram method (ADM)—The traditional or first method for representing a logic network by identifying an activity as an arrow with circles (or events), noting its beginning and ending.

as late as possible (ALAP)—In critical path method (CPM) and project evaluation and review technique (PERT), designates a task that should start as late as possible, using all the free-float time available. See also *critical path method* and *project evaluation and review technique.*

as soon as possible (ASAP)—In critical path method (CPM) and project evaluation and review technique (PERT), indicates a task that should start as soon as possible, given the start date of the project and its predecessor tasks. See also *critical path method* and *project evaluation and review technique.*

average job estimate—The estimated average time spent on a troubleshooting or repair job.

backlog age—A measurement of the number of work orders in the backlog that can be completed within their priority period, and the number that has missed the deadline.

backlog hours—The number of direct labor hours identified by work orders currently on hand.

balance sheet—An accounting of the raw data used to generate performance indices and graphic presentations in the weekly report of the activities. Also shows the assets and liabilities for a business.

baseline—The original project plan in project, shutdown, outage, and turnaround planning (including time schedule, as well as resource and cost allocations). Used for comparison against actuals to track and analyze a project's progress.

battery limits—Area around equipment, which is usually confined.

benchmark (in MTM methods)—A short description of a job and an estimate of the labor hours required to complete it. The description and estimate are usually laid out on a spreadsheet by job type or craft, and in order of labor hours required. See also *methods time measurement.*

benchmark job—A short description of a job and an estimate of the labor hours required to complete it. The description and estimate are usually laid out on a spreadsheet by job type or craft, and in order of labor hours required.

book value of a facility—The purchase and construction cost of buildings and equipment in a facility, less the total depreciation of those assets.

budget estimate—An estimate to help inform the client of the project's financial scope and whether the project is within a budget.

business practice re-engineering (BPR)—A process that manages quality and eliminates non–value-added activities.

calibration—Confirmation or correction of the accuracy of critical indicators, control instruments, or final elements.

classic curves—The costs associated with preventive maintenance (PM) and the costs associated with production. See also *preventive maintenance.*

computerized maintenance management system (CMMS)—A computer system that tracks the relationship between work orders and associated records.

conceptual estimate—An estimate (based on the available design information) that uses rough tables and charts to price equipment and labor.

condition priority number—A number (based on the condition and priority) that the originator (of the work) should assign to a job.

contrived emergency—Event that occurs when the originator (of the work) believes that the job will not be started unless it has a high priority. This is not a real emergency.

corrective work—The maintenance work performed during maintenance custody to replace worn parts, adjust loose equipment, prevent a major failure, and return the equipment to nameplate condition. It can also refer to the plannable repairs on installed spares during production custody.

CPM—See *critical path method*.

crash time—The shortest duration of the project.

critical path—The sequence of tasks or activities whose schedules and durations directly affect the date of overall project completion. Also, the series of jobs in a critical path network that make up the longest path to the completion of a project. Alternatively can be the sequential list of jobs from the beginning to the end of a critical path network, having no float. See also *critical path method* and *float*.

critical path method (CPM)—A system of project scheduling used to identify the sequence of activities and milestones required to complete a project.

critical path network—Used in critical path method (CPM), the complete diagram of all the activities and milestones required to complete a project. See also *critical path method*.

custody plan—An intelligent schedule for operations uptime and maintenance downtime. Also identifies the custody of the equipment by both the operations department and the maintenance department.

design review estimate—An estimate with as few temporary figures as possible. Also provides the sources of information (such as labor rates and crew size).

discretionary PM requirements—The activities performed on equipment or at a facility that are based on the need to reduce downtime, minimize costly equipment damage, or ensure personnel safety. See also *preventive maintenance.*

displacement —A common measurement on slow-speed equipment (operating at 600 rpm and below). Measured in mils.

dollar estimates—A type of estimation used in the control of the funds available for maintenance. The three common reasons for developing dollar estimates are making contract (or buy) decisions, approvals, and budgeting.

earliest completion time—The calculation of the earliest possible finish time of each activity. Calculated by starting at the beginning event and adding up subsequent activity times.

earned value—A performance measure calculated by multiplying a task's planned cost by the percentage of work completed. Required in all projects for the U.S. government.

effectiveness factor—Determined by dividing the estimated hours (not including delay) of jobs worked the past week, month, or year by the actual hours worked.

elapsed time—The total time required to complete a job or task. Elapsed time is equal to the labor hours if only one person is assigned to the job. If more than one person is assigned to the job, the elapsed time will be less than the labor hours.

emergency work—Referred to as emergent work in nuclear power plants. An emergency is a classification of maintenance work that results from breakdown of critical equipment, usually requiring immediate action. An emergency can also be defined as any job that displaces work on a schedule.

engineered time standard—A standard for a method of performing a task by measuring several people completing that task.

equipment priority number—A number assigned to indicate the relative importance of each location in the process or system. The higher the number assigned to a location, the lower the equipment priority number.

era of business management theories—Period from the 1960s to the present when several management-related programs were first introduced.

event—Used in critical path method (CPM), arrow diagram to describe the starting point of one activity and the beginning point of another activity. An event will be indicated in a critical path network diagram by a number enclosed in a circle. See also *critical path method* and *critical path network*.

ferrographic analysis—Inspection of the physical size, shape, and size distribution of wear metals in lubricating oil.

float—The amount of time a noncritical task can be delayed before it influences another task's schedule. Also called *slack*.

four Ts of correction—Time, target, tools, and training that assist an employee with the problem relating to vibration correction.

Gantt chart—A graphical representation of a project schedule, which shows each task as a bar whose length is proportional to its duration. Many programs include a spreadsheet section to the left of the Gantt chart to display a selection of project data.

generic maintenance—Work that does not change much from industry to industry, in which the job steps do not change with the size of the equipment.

grant of authority—Assigned grant to individuals in a company that allows them to make purchases or authorize maintenance work only up to their assigned amount.

hertz—Unit of frequency. A periodic oscillation has a frequency of n hertz (Hz) if, in 1 second, it goes through n cycles (cycles per second).

histogram—In critical path method (CPM) and project evaluation and review technique (PERT), a bar chart that shows resource workloads by time period. See also *critical path method* and *project evaluation and review technique*.

hour estimates—An estimate based on the number of hours used in the control of a limited labor resource. The most common reasons for developing hour estimates are scheduling, backlog evaluation, critical path method (CPM) and project evaluation and review technique (PERT) data. See also *critical path method* and *project evaluation and review technique*.

index—A tool to help predict future activity, or to compare current activity to a standard. An index identifies negative trends before they become too costly and highlights the success or failure of programs recently instituted by management. The value of an index is related to the cost of obtaining it.

inspection—Using the basic senses of sight, sound, and touch on operating machinery. Moreover, it can be improved by adding gauges or meters to the equipment being inspected.

just-in-time (JIT)—In the broad sense, an approach to achieving excellence in manufacturing, based on the continuing elimination of waste (waste being considered to be anything that does not add value to the product). Benefits include the elimination of all unnecessary inventories. The elimination of much of the raw material and work-in-process inventories forces suppliers and each production step to produce impeccable quality, since without buffer inventories, poor quality will cause the production process to stop. Other benefits include less space required (no inventory storage), faster throughput times, increased productivity, and lower costs. In the narrow sense, JIT refers to the movement of material at the necessary place at the necessary time. The implication is that each operation is closely synchronized with the previous and subsequent ones to make that possible.

key work order—The primary craft (or the craft) that will complete most of the job.

kit—The components of an assembly that have been pulled from stock and packaged for use. Examples are components from which an assembly is to be produced, or a group of gaskets readied for an engine overhaul.

labor hours—The total hours worked by all individuals on a job or task. Formerly referred to as *man-hours*.

lag—In critical path method (CPM) and project evaluation and review technique (PERT), the amount of time between the finish of a predecessor task and the start of a successor task. See also *critical path method* and *project evaluation and review technique*.

late start—The additional time that is spent after an official break is completed. An example is the time spent traveling back to the job after the break is finished.

latest completion time—The latest possible time that each activity can be finished without increasing the length of the project. Determined by subtracting activity times from the total project elapsed time at the end of the critical path network until the first activity is reached. See also *critical path network*.

lead—In critical path method (CPM) and project evaluation and review technique (PERT), the amount of time that a successor task is permitted to start before its predecessor is finished. See also *critical path method* and *project evaluation and review technique*.

learning curve—A graphical representation of the fact that, in repetitive activities, there is a constant and predictable rate of productivity improvement each time the number of units produced is doubled. Rates will vary with the activity or product being produced. Nevertheless, they will exhibit the same characteristics of constancy and predictability with the doubling of repetitions. As a purchasing tool, this knowledge is particularly useful in negotiating with suppliers for custom-made products.

load leveling—The distribution of resources so that constraints on the resources are not violated.

lubrication analysis—An analysis of lubricating oil to detect bearing failure, inspect the wear particles in the oil, and monitor certain equipment.

maintenance by default—Equipment repaired as it fails, usually on an emergency basis.

maintenance by plan—Using forethought to determine what level of maintenance is required.

management by objective (MBO)—A theory of business management whereby the business achieves specific objectives under the control of managers.

management by results (MBR)—A theory of business management whereby management looks at the results of the past as an indication of the results of the future.

mandatory PM requirements—Activities performed on equipment or facilities as required by law or contract. See also *preventive maintenance*.

man-hours—See *labor hours*.

methods time measurement (MTM)—The development of job steps based on the basic movements of a human's hands, feet, eyes, and body. Times associated with these movements are measured down to the nearest 0.00001 hour.

milestone—A project event that represents a checkpoint, a major accomplishment, or a measurable goal; also, a significant point in the development of a project or job. Commonly used in critical path method (CPM). See also *critical path method*.

most-common case estimate—The estimate of the most-common time spent on a troubleshooting or repair job.

moving average—The average of data in a window over a set period (such as 3 months, 6 months, or 1 year). Used to smooth out major swings in rates and to better evaluate trends.

multiskilling—The process of training maintenance employees in specific skills that cross the traditional trade or craft lines, and then ensuring that the work is performed.

nameplate—A plaque giving the manufacturer's name and the rating of the machine.

negative float (negative slack)—In critical path method (CPM) and project evaluation and review technique (PERT), unscheduled delay before an actual task starts. This is time that must be recovered if the project is not to be delayed. See also *critical path method* and *project evaluation and review technique.*

performance indices—Defined performance-related reliability measurements.

PERT—See *project evaluation and review technique.*

PERT chart—A graphical depiction of task dependencies resembling a flowchart. Dependencies are indicated by connecting lines or arrows that show the work flow. Also called a *network diagram.*

planning—The allocation of needed resources and the sequence in which they are needed to allow an essential activity to be performed in the shortest time or at the least cost.

planning thought process—A method that uses the abilities of a planner to the fullest to establish an estimate of time required for the job, define understandable steps to complete the job, and identify the materials, parts, tools, and equipment required for the job.

plug—A temporary figure that is used until a better estimate can be developed. Sometimes referred to as a *plug-in.*

precedent diagram method (PDM)—A method for dispensing with event identification and instead using a box to represent the activity. Lines (drawn from left to right) represent the interdependencies of different activities.

precedent logic—The method of putting activities in sequential order.

predecessor—In critical path method (CPM) and project evaluation and review technique (PERT), the task that must be started or completed first in a dependency relationship between two tasks. See also *critical path method* and *project evaluation and review technique.*

predesign estimate—An estimate whereby the design information has not been improved, but a client may want additional background information from potential construction firms who may be awarded the project.

predictive maintenance (PDM)—A program that compares test measurements to established engineering limits in order to determine the need for corrective work. The limits are set to ensure that sufficient time is available for repair and to prevent an emergency shutdown of the equipment.

prefabrication—The maintenance work during production custody that includes the rebuild of equipment in the shop, laying out pipe for future replacement, and setup of equipment prior to a planned shutdown.

preventive maintenance (PM)—The maintenance work performed during maintenance custody on equipment through manufacturer-recommended rebuilds or rebuilds required because of predictable wear; also, the maintenance work performed during production custody on equipment (including adjustments, lubrication, tests, inspections, and calibrations).

prioritization—A process that begins as requests for maintenance work are received, and breaks down the requests into three basic stages: classification, requested completion date, and schedule priorities.

project evaluation and review technique (PERT)—A method of reviewing projects similar to the critical path method, except that it also incorporates the uncertainty associated with completing each task in the project. A PERT diagram can now also (erroneously) refer to the network diagram developed through critical path method (CPM) in many computerized project programs. This new definition has had growing acceptance because of the proliferation of project programs, even though it has little to do with the original intent of the method.

purchasing agent—The person authorized by the company to purchase goods and services for the company; one to whom authority has been delegated to enter into contracts on behalf of the firm.

real emergency—A situation in which something is broken and must be fixed. These emergencies consist of two types: those that are foreseeable and those that are not.

rebuild—Checking of critical dimensions and replacing of worn parts. Requires equipment downtime.

reliability-centered maintenance (RCM)—The process to refocus maintenance on overall equipment reliability while reducing the cost of needless maintenance services.

replacement PM—Involves the periodic replacement of disposable components and can also be a periodic change-out of equipment or components in anticipation of failure. See also *preventive maintenance.*

resource—Any person, group of people, piece of equipment, or material used in accomplishing a task or job.

resource-driven—Task durations determined by a project program based on the number and allocation of resources, rather than the time available. Both tasks and entire projects can be resource-driven.

resource leveling—The process of resolving resource conflicts.

safety or environmental modifier—An additional modifier to equipment and condition priority numbers in relation to safety or environmental compliance. See also *condition priority number.*

safety stock—That quantity of inventory (above normal usage) that is kept as protection against uncertainty of demand or of supply availability.

scatter diagram—A method of graphing data to show the actual distribution of the data.

scheduling—The assignment of many planned jobs into a defined period of time in order to optimize the use of the resources within their constraints. Effective scheduling cannot be accomplished without planning.

shutdown—Down period used for inspections in organizations; also referred to as *shut-in, downturn, turnaround,* or *outage.*

single-line diagram—A form of functional shorthand whereby a single line is used to represent the three phases of an ac power-distribution system, and a three-phase device (such as a motor starter) is represented by a part associated with only one phase.

slack—See *float.*

square-foot cost estimating—An estimating method used to prepare budget or preliminary estimates for construction projects.

standard labor rate—All the costs associated with having the worker on the payroll, not just the hourly pay. (This includes other items such as vacation pay, holiday pay, and other benefits.)

statistical process control (SPC)—A quality control methodology that focuses on continuous monitoring and plotting on charts of quality levels during the production process itself, rather than on postproduction inspection of the items produced. The intent is to produce no defective items. By recording quality results (at frequent regular intervals) on a chart, trends toward defective production can be spotted. By stopping the process before it drifts out of control, and by making necessary corrections, no defective parts will be produced. The machine operator then becomes his or her own quality inspector.

stores withdrawal slips—Documents that contain pertinent information relating to the withdrawal of stock from the storeroom.

subproject—A group of activities treated as a single task in a master project schedule. Subprojects are a way of working with multiple projects by keeping all the data in one file rather than independent files.

successor—A task that must await the start or completion of another in a dependency relationship between two tasks.

take-off—Bill of material or document that records the dimensions and quantities to be included in the estimate.

therblig—A basic element of the human body's movement of a given manual operation.

three-phase design—A design for ac motors using more than 10 horsepower that results in a tendency for the motors to vibrate less because they require at least one cable for each phase. See also *vibration analysis*.

time domain logic network—Plotting of activities to show their elapsed time, as well as the dependencies with prior and later activities. Combines the advantages of the critical path method with the ease of viewing a project in Gantt chart form. See also *critical path method* and *Gantt chart*.

time measurement unit (TMU)—The unit of measurement that records and times the elemental activities of the worker. One-hundred-thousandth of an hour is equal to 100,000 TMUs (0.00001 hour = 0.0006 minute = 0.036 second, or 1 hour = 100,000 TMU).

time-and-motion study—A study to determine the most efficient method (in terms of time and effort) of a specific task to increase the profits of a business.

total productive maintenance (TPM)—A theory of business management that extended the idea of autonomy and suggested that production-line workers could also perform maintenance on equipment.

total quality management (TQM)—A management theory that focuses on the reduction of defects in the process at every point in the process to reach the ultimate goal of zero defects.

unit cost—The value assigned to a single unit of work. The total cost is determined when the unit cost is multiplied by a quantity.

vibration analysis—An analysis of vibration in rotating machinery to discover problems such as misalignment, imbalance, bearing damage, and looseness.

work order—A formal means of requesting maintenance services, making it the focus of most maintenance records; provides a financial structure to the work that the maintenance department performs.

work package—A folder (or group of folders) that includes all the necessary paperwork and reference material required to complete a job. The work order, written procedure, specification sheets, tool lists, parts lists, sketches, prints, permits, and equipment manuals should be included.

work request classification—A work request can be assigned a classification after first generated.

workable backlog index—Calculated from a total of all the work that is ready because parts and other material are on-site.

worst-case estimate—The estimate of the worst-case amount of time spent on a troubleshooting or repair job.

Index